Oracle Digital Assistant

A Guide to Enterprise-Grade Chatbots

Luc Bors
Ardhendu Samajdwer
Mascha van Oosterhout

Apress®

Oracle Digital Assistant: A Guide to Enterprise-Grade Chatbots

Luc Bors
Houten, The Netherlands

Ardhendu Samajdwer
Bilthoven, The Netherlands

Mascha van Oosterhout
Bilthoven, The Netherlands

ISBN-13 (pbk): 978-1-4842-5421-9
https://doi.org/10.1007/978-1-4842-5422-6

ISBN-13 (electronic): 978-1-4842-5422-6

Managing Director, Apress Media LLC: Welmoed Spahr
Acquisitions Editor: Jonathan Gennick
Development Editor: Laura Berendson
Coordinating Editor: Jill Balzano

Cover image designed by Freepik (www.freepik.com)

Distributed to the book trade worldwide by Springer Science+Business Media New York, 233 Spring Street, 6th Floor, New York, NY 10013. Phone 1-800-SPRINGER, fax (201) 348-4505, e-mail orders-ny@springer-sbm.com, or visit www.springeronline.com. Apress Media, LLC is a California LLC and the sole member (owner) is Springer Science + Business Media Finance Inc (SSBM Finance Inc). SSBM Finance Inc is a **Delaware** corporation.

For information on translations, please e-mail rights@apress.com, or visit http://www.apress.com/rights-permissions.

Apress titles may be purchased in bulk for academic, corporate, or promotional use. eBook versions and licenses are also available for most titles. For more information, reference our Print and eBook Bulk Sales web page at http://www.apress.com/bulk-sales.

Any source code or other supplementary material referenced by the author in this book is available to readers on GitHub via the book's product page, located at www.apress.com/9781484254219. For more detailed information, please visit http://www.apress.com/source-code.

Printed on acid-free paper

"The only way to do great work is to love what you do."
—Steve Jobs

Every accomplishment starts with the decision to try.

Table of Contents

v

About the Authors

Luc Bors is Partner and Technical Director at eProseed and member of the global eProseed CTO Office. He has over 20 years of experience in IT. He is a certified specialist in Oracle Application Development Framework (ADF), Oracle MAF, Oracle MCS, Oracle JET, and Oracle Digital Assistant (ODA). He is very interested in new technologies such as chatbots, Digital Assistants (DAs), and IoT.

Luc was promoted to Oracle ACE Director in 2015 because of his experience and knowledge and for his willingness to share with the community. In 2017 he was among the first to become an Oracle Developer Champion, currently known as Oracle Groundbreaker Ambassador. He is the author of the book *Oracle Mobile Application Framework Developer Guide*, is a speaker at international conferences, and regularly publishes articles on his blog and in several printed and digital magazines.

Ardhendu Samajdwer is Senior R&D Architect and UI Consultant at eProseed. He has over 12 years of experience in the IT industry and is Oracle Cloud Platform Enterprise Mobile 2018 Certified Associate Developer. He has been part of development and maintenance of various projects involving migration and implementation of enterprise-level applications. Ardhendu is currently working on Oracle Platform as a Service (PaaS) offerings such as Visual Builder Cloud, Mobile Cloud, and Chatbots in addition to front-end technologies including Oracle JET and React. Lately, he has started making appearances, as a speaker, in OUG conferences.

 **Mascha van Oosterhout** is User Experience Consultant at eProseed. She has a master's degree in industrial design engineering from the Faculty of Human Factors and Design at Delft University of Technology in the Netherlands. She has been a user experience design (UXD) consultant for more than 20 years. She has worked in both private sector companies and public sector organizations and in support of start-up initiatives.

Mascha is experienced in designing, reviewing, and testing graphically clear and user-friendly interfaces. She has passion for putting the end user first and foremost. She gathers user requirements and then specifies interaction design by means of wireframes and graphical layouts. Recently she has increased focus on chatbot dialogs because she is convinced that, in many contexts, conversations are a more natural way to interact than through a graphical interface. She is an expert in user research techniques such as persona modeling, formulating user scenarios, interviewing stakeholders, doing cart sorting research, and usability testing.

About the Technical Reviewer

Frank Nimphius is Master Principal Product Manager at Oracle Corporation. He joined Oracle Product Management more than 20 years ago and has been involved in the development of several Oracle products, including Oracle Forms, Oracle JDeveloper, the Oracle Application Development Framework (ADF), and Oracle Mobile Hub (OMH). In his current product management position, Frank works in the Oracle Digital Assistant (ODA) group that builds Oracle's strategic conversational AI platform. For the last 20+ years at Oracle, Frank enjoyed the many opportunities the product management role offered and became a book author, technical trainer, speaker at international conferences, author of technical collateral and trainings, Java and JavaScript developer, and blogger.

Acknowledgments

This is from Luc:

Never say never again; this is my lesson learned. When I finished my first book some 4 years ago, I was convinced that I would never ever write a book again. Somehow that changed when Oracle released their Digital Assistant. I immediately was and still am convinced of the power of this product and the added value that it provides for our customers.

My urge to share knowledge and the passion for new technologies took over and made me change my mind and tricked me into writing again.

Thanks to Lonneke Dikmans who expressed her support to this project and suggested that I should set up a team of coauthors to make this happen. I could never have done this without the help of Mascha and Ardhendu who added a tremendous amount of effort, enthusiasm, and humor to this project. Thanks to you both for that! Without you, this book would have never been written.

Special credits go to the team at Oracle who helped with the technical review, but also provided early access at some point and guidance in technical challenges. Thanks, Frank, Grant, Martin, Rohit, and all others who helped at some point, even if you don't know that you did.

Finally, hugs and kisses for my wife, Judith, and kids, Olivier and Lisanne, who again had to live with me during the process of writing. Must have been difficult, but you never stopped supporting me. Thanks!

This is from Ardhendu:

Am I dreaming or is it really happening?! Tough for me to realize myself as an author today. "Thanks" would be too small to say to you, **Luc**. It would have been impossible for me without your guidance and mentoring. And **Mascha**, for always cheering me up and guiding me. I want to extend my thanks to **Lonneke Dikmans** and **Ronald van Luttikhuizen** for their constant encouragement, support, and guidance during the course of writing this book.

ACKNOWLEDGMENTS

Most importantly, I want to thank my parents, without whom I could not have been where I am today. Finally, I want to thank my lovely wife and my little princess; you both have always been my source of inspiration.

Special thanks to our publishers for providing us this platform.

This is from Mascha:

Never thought I would really do this, being a coauthor and publishing a book together with my two great eProseed colleagues Luc and Ardhendu. I would not have been able to achieve this without them. They have offered me, as a first-time author, the best guidance, feedback, and suggestions along the way.

Thank you, Luc, for asking me to cooperate in making this happen; without you, I would never have enjoyed this opportunity of being able to write about the other side of coding a digital assistant, namely, the user side. Being a user experience designer, I always put the user first and foremost. We **sell** and **develop** Digital Assistants for our **customers**, but we **create** them for their **users**.

I also want to thank Lonneke Dikmans and Ronald van Luttikhuizen for encouraging and supporting me and many times expressing their enthusiasm about the book.

I thank Hans Kemp for enabling me to teach young UX professionals about the conversational user experience design method which made it possible for me to refine my chapters.

I am immeasurably grateful to my dear husband for telling anyone who wants to hear it that his wife, together with two of her colleagues, wrote a book – her debut – that is published for real and therefore is for sale ☺.

Finally, I want to thank my beloved parents, for being enthusiastic and proud of me, for believing in me, and for always reassuring me by saying "Als iemand kan schrijven, ben jij het."

Introduction

Chatbots, or nowadays mostly called Digital Assistants, found their way into the enterprise. They assist users to execute administrative tasks in an interactive way, making these tasks simpler and less time-consuming. Throughout the book, you will learn to understand the core concepts of Digital Assistants.

This book provides you with an on-ramp to the development of enterprise-grade chatbots and digital assistants. It is based on real-life experience and explains everything you need to know to start building your own digital assistant (DA) using Oracle technologies. By reading this book, you will become familiar with the concepts involved in DA development from both a user experience and a technical point of view. You will learn to create Digital Assistants using Oracle technologies, including the Oracle Digital Assistant Cloud.

In the first part of the book, you will learn the basic principles of digital assistant – AKA chatbot – technology. After this you will be guided through the steps involved in designing a Digital Assistant, including how to make sure that the user receives a satisfying experience when using the assistant. In the second part of the book, you will learn how to implement the digital assistant that was designed in the first part. You'll begin with a basic implementation, later enhancing that implementation with multilanguage support, Q and A (QnA), and Webviews. The final part of the book adds a deep dive into custom component development, sentiment analyses, and speech.

This book is intended for designers and developers who want to implement digital assistants using Oracle technologies and cloud platform. This book is ideal for readers new to creating digital assistants and covers aspects of design, including user experience design, before moving into the technical implementation. Readers experienced in creating digital assistants on other platforms will find the book useful for making the transition to Oracle technologies and the Oracle Digital Assistant Cloud.

> **Chapter 1.** This chapter introduces Oracle Digital Assistant as a platform for developing apps for natural conversational interfaces. The platform makes it easy to build sophisticated digital assistants or simple chatbots that can connect and extend multiple backend systems, like Oracle ERP, HCM, and CX.

Chapter 2. In this chapter, you will find answers to key questions that you face when designing Digital Assistants. What is the business pain we are trying to solve? What is the existing journey users take and how can it be improved? What is the appropriate channel to reach the users and how can it best be used? What is the conversation between the bot and user and the dialog option and how can the dialog be enhanced to use the channel's properties, media, and facilities to improve the user interactivity? What are the key vocabulary and entities exchanged in the dialog and needed to access the backend systems? Many, many questions that need to be answered before the actual implementation can be achieved.

Chapter 3. To make sense of the technical implementation of the Digital Assistant, in this chapter, you will be introduced to Travvy, the Digital Assistant for Extreme Hiking Holidays. Using a step-by-step approach, you will get acquainted with Travvy so you know all the ins and outs in order to understand all the steps of the technical implementation.

Chapter 4. In this chapter, you will learn how Oracle Digital Assistant allows developers to sit with designers to enter the initial flow into the Digital Assistant Cloud. This will speed up the transition between design and technical implementation, as the designer actually can see how the DA behaves and how the flow is executed.

Chapter 5. This chapter explains the areas involved in the technical implementation, such as flow, intents, and entities. Together these form the heart and body of your digital assistant. You will learn to implement and understand this and how to use the training facilities in order to make your DA understand what the user means.

Chapter 6. In this chapter, you will introduce your Digital Assistant to the user. You will learn how to configure channels. The DA can run on any messaging service that supports webhooks, calls that allow real-time messaging without polling

Chapter 7. This is where you learn how to implement multilanguage support: You can use multiple-language approach or single-language approach. The latter is what is used by Oracle Digital Assistant, extended with the use of translation services.

Chapter 8. Natural language conversations are, by their very nature, free-flowing. But they may not always be the best way for your bot to collect information from its users. For example, some situations, like entering credit card or passport details, require users to enter specific information (and enter it precisely). To help your bot's users to enter this type of information easily, your bot can call an external application, which provides forms with labels, options, choices, check boxes, data fields, and other UI elements. You will learn how to achieve this by using Webview components.

Chapter 9. Frequently asked questions (FAQs), by their very nature, are common questions that are simply looking for an answer: "What are your opening times?", "Can I use my credit card?", "How many times a week do you deliver?" These FAQs (or QnAs) often already exist in a company. In this chapter, you will learn how to bring them into your bot.

Chapter 10. You will learn how to build Custom Components. Whenever the Digital Assistant needs a specific action that's outside of the functions provided by the built-in components, such as returning backend data or implementing business logic, we need to use a custom component. These components are specific to the use case, so they need to be specifically built.

Chapter 11. The reasoning behind using sentiment analysis is human psychology. When they feel happy or neutral, people tend to take bad news or frustration in a more accepting way. By understanding context right from the beginning, a chatbot can select the best course of action and apply very different patterns. You will learn how to add sentiment analyses to your Digital assistant, and in addition to that, you will learn how to implement speech as a channel.

We have created a weblog dedicated to this book and the Digital Assistant in general:

`https://oda-book.blogspot.com/`

This weblog will be used to post additional content and other articles that can be used to build great Digital Assistants.

No prior knowledge of chatbot technology is assumed. Everything you need to know to become a master is contained in this book. We hope you enjoy this book!

PART I

Foundations

CHAPTER 1

Introduction to Oracle Digital Assistant

Over the past decades, the size of computers has changed from huge warehouse-sized supercomputers to tiny pocketsize smartphones. The way people interact with computers or use programs has changed too. From client server or browser based on desktop and laptop to the use of apps on tablet and smartphone. Over the last couple of years, the use of apps has changed drastically. Where originally there used to be "one app for each task," nowadays more and more people want to have "one app for all tasks." The use of messaging apps such as WeChat, Snapchat, Skype, Slack, and Facebook Messenger has exploded. Besides that, virtual private assistants such as Google Home, Amazon Echo, Apple Siri, and Microsoft Cortana found their way into people's lives. Let's have a look at how this happened and how Oracle took the opportunity to fill a gap between consumer use and enterprise use.

A Brief History of Chatbots

The history of chatbots started about 70 years ago when Alan Turing[1] stated that a truly intelligent machine is one that cannot be distinguished from a human in a text-only conversation. Eventually this led to the "Turing Test" which can be used to prove whether a machine is truly exhibiting intelligent behavior or not. In his 1950 paper "Computing Machinery and Intelligence," he laid the foundation of what we now know as chatbots or Digital Assistants.

---

[1]**Alan Turing is** often called the father of modern computing. He was a brilliant mathematician and logician. He developed the idea of the modern computer and artificial intelligence. During the Second World War, he worked for the English government breaking the enemy's codes, and Churchill said he shortened the war by 2 years.

© Luc Bors, Ardhendu Samajdwer, Mascha van Oosterhout 2020
L. Bors et al., *Oracle Digital Assistant*, https://doi.org/10.1007/978-1-4842-5422-6_1

About 15 years after Turing, MIT AI Laboratory created ELIZA. This was a natural language processing computer program, able to simulate human conversations, based on scripted responses. Where many early users were convinced of ELIZA's intelligence and understanding, in fact ELIZA was not able to have a conversation with true understanding, and it did not take a very long time before it became very obvious that you are talking to a machine

Shortly after ELIZA, in 1972 PARRY was created at Stanford University. PARRY was the first "chatbot" that actually had a conversational strategy. During the first International Conference on Computer Communications in 1972, PARRY and ELIZA were set up to talk to each other.

From ELIZA and PARRY to the next milestone in "chatbot" history, it took almost two decades. In the late 1980s, Jabberwacky was developed. Jabberwacky (still online at www.jabberwacky.com) aims to simulate natural human chat in an interesting, entertaining, and humorous manner. Jabberwacky is the first "chatbot" that uses artificial intelligence and as such is able to learn. It learns from the conversations that it has had in the past.

At the end of the last century and the beginning of the millennium, other "chatbots" made their appearance such as Dr. Sbaitso, ALICE, and SmarterChild. ALICE (Artificial Linguistic Internet Computer Entity) was based on ELIZA. Even though ALICE can set up a conversation with a user, based on user input and pattern recognition, ALICE has never been able to pass the Turing Test.

The next-generation chatbots included bots from big vendors such as IBM Watson (2006), Apple Siri (2010), Google Now (2012), Amazon Alexa (2015), and Microsoft Cortana (2015). In 2016, Facebook introduced bots for Facebook Messenger that by the end of the year had over 30,000 bots available.

Chatbots are generally used as interfaces, such as extracting product details. Chatbots are task oriented, whereas digital assistants are user focused. Digital assistants provide the ability to combine different chatbot tasks to a single conversation. These assistants can really assist users with tasks such as reminding you of meetings, managing your to-do lists, and so on. When a chatbot is asked to provide such virtual assistance, they usually get confused and ultimately keep asking the same questions for clarification. It is very important to understand that artificial intelligence is not all that it takes to build chatbots. Other skills, including UX and conversation design, play an important role. Bad design can even make a digital assistant fail. Even though chatbots and digital assistants are both considered conversational interfaces, they are very different. One of the biggest differences can be found in how the two maintain conversational flow. When interacting with chatbots, if you interrupt the conversation in between, the chatbot will most probably fail to remember the context of the interaction, whereas virtual assistants use dynamic conversation flow techniques, so they can understand human intent and keep the flow going.

The following example shows how a well-designed assistant can cope with context change from booking ticket to checking balance:

"I like to **book a ticket** for the Taylor Swift concert in Amsterdam."

– Sure, any preference for where you want to enjoy the gig? We have category A for 84 Euro, category B for 60 Euro, and category C for 35 Euro.

"**How much money** do I have?"

– Which account do you want to know your balance for? A) Savings. B) Credit card. C) Checking account.

"C, checking."

– Well, you have 1500 Euro available.

"Two tickets, category A please."

– Sure. Are you interested in a Taylor Swift merchandise too? We have them on account today.

Most Digital Assistants nowadays go beyond providing simple "request and response" features. They are changing big time the way that brands engage with their customers.

Oracle Digital Assistant

The majority of the aforementioned chatbots were aiming for personal use. Although they could also be used in enterprise solutions, looking at requirements, there were many gaps between what the solutions offer and what the enterprise needs.

Bridging the gap between personal Digital Assistants and enterprise-grade Digital Assistants is exactly where Oracle aims to be, and by the end of 2016, Oracle released their first version of Oracle Digital Assistant. At that time, it was called Oracle Chatbot, and it was part of the Oracle Mobile Cloud Service. Some 2 years later, Oracle released

Oracle Digital Assistant (ODA) cloud service, a product dedicated to developing and running digital assistants. Digital assistants are virtual personal assistants that understand natural language when interacting with users. Typically, simple chatbots resolve users' intent and help them to complete a simple task. Oracle Digital Assistant goes one step beyond that. Each digital assistant contains a collection of specialized skills. The assistants can be trained to use these multiple different "skills." These skills can have their own backend domain integrations and can be used in the course of an across-skill user conversation to work with these backends. They will evaluate user input and invoke the appropriate skill to start or continue the conversation.

Through this approach, everything a user needs to do can be done in a single conversation, without the need to work with multiple chatbots. Users can also switch in the middle of a conversation, and the Digital Assistant will still understand the context of the users' questions. Sharing of context between the skills is key to success and very important in providing a good user experience. Think of something like someone's location, mood, or needs in a particular moment. These attributes can impact what messages the user will be receptive to and, even more critically, when and where.

Oracle Digital Assistant is also capable of initiating a conversation with a user, based on scheduled events that are received from external applications. This is called Application-Initiated Conversation (AIC). The Oracle Digital Assistant can use content of the application's event message to begin a conversation at a predetermined state in the dialog. This is something that differentiates Oracle Digital Assistant from other chatbots, which typically only "speak when spoken to."

Oracle Digital Assistant Core Components

To get a clear understanding of Oracle Digital Assistant (ODA), you need to know its high-level architecture. Oracle Digital Assistant consists of a set of components working together to enable the development and use of enterprise-grade digital assistants. In this section, you will learn what these components are and what their main purpose is.

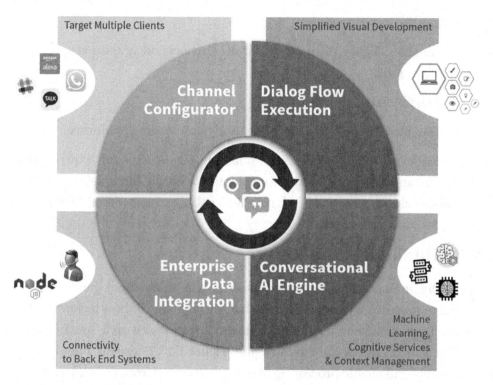

Figure 1-1. *Oracle Digital Assistant – core components*

The Oracle Digital Assistant is a Platform as a Service (PaaS) that enables you to create digital assistants and to expose these to many different interfaces, called channels. Oracle Digital Assistant allows you to create both **Digital Assistants** and **Skills**.

Digital assistants can help users to do multiple tasks in natural language conversations. Each digital assistant typically has its own set of skills.

Skills can be seen as conversational workhorses that can be used with one or many digital assistants. Think of tasks such as ordering flowers in a specific shop or checking the balance on a bank account. With Oracle Digital Assistant, the use of a skill is not limited to a single digital assistant.

Implementation of Skills and Digital Assistants relies on four core components (Figure 1-1) working together:

- The conversational AI engine

  - Enables the Digital Assistant to work with user input

- The dialog flow

  - Defines the interactions that users can have with a specific skill

7

- Enterprise data integration

  - Enables Digital Assistants to connect to backend systems via their skills

- The channel configurator

  - Enables Digital Assistants to be used in messaging platforms

Before we dive into the development of skills and digital assistants, you need to get some understanding of some of these concepts, which will be explained in the remainder of this chapter. We will first have a look at the conversational AI engine.

Conversational AI Engine

When using Oracle Digital Assistant, you do not have to worry about the technologies that are used to process and understand the natural language or how to work with user input. Oracle Digital Assistant uses different technologies based on neural networks to implement its own conversational AI engine (Figure 1-2). These use linguistic and language modeling to enable processing of natural language from the end user.

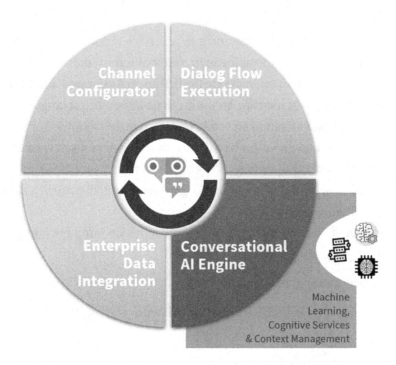

Figure 1-2. *Oracle Digital Assistant – conversational AI engine*

With this in place, the developer can focus on the user conversations instead of having to know all the nitty-gritty details of these underlying algorithms.

Both intents and entities are common NLP (natural language processing) concepts. NLP is the science of extracting the intention of text and relevant information from text.

Intents are groups of tasks or actions that a user expects your skill to perform for them. Intents usually include verbs and nouns such as "give quote," "get dates," and "find trip." They allow your skill to understand what the user wants it to do. In other words, they can determine the users' intent. For instance, a "FindTrip" intent can relate to a direct instruction, such as *I want to find a trip*, but also to other requests, like *I'm really in for a short holiday*, both of which are **utterances** for the same intent.

Whereas intents map words and phrases to a specific action, **entities** add context to the intent itself. They are key identifiers for pieces of user input which enable a skill to fulfill a task. Basically, entities are words that modify the intents (big, modern, outdoor, short, best available). Entities can be divided into user-created entities and system entities. All of the aforementioned concepts will be discussed in more detail throughout this book.

Dialog Flow Execution

The next component to introduce is the **dialog flow** (Figure 1-3). A dialog flow defines the possible interactions users can have with a skill. It describes how a skill responds and behaves according to user input.

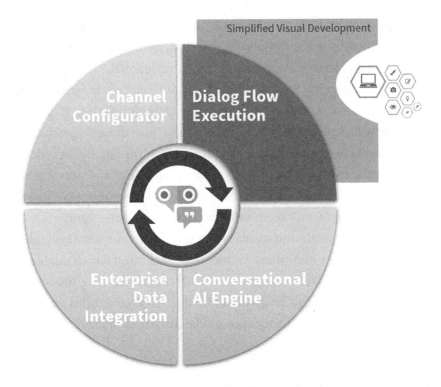

Figure 1-3. *Oracle Digital Assistant – dialog flow execution*

Oracle Digital Assistant enables you to define a context-aware conversational dialog. To have a context-aware dialog is very important as an end user will not necessarily stick to the matter and can potentially branch off into different states and context during a conversation. For example, if a user wants to buy flowers, but before he can continue to the payment, he has to check the balance on his account.

At some point in the flow, they would instruct the assistant to "Pay the flowers." The response of the assistant could be "from which account." The user would pick "Checking Account" but actually has no clue how much money there is in that account. Then the user would switch context to ask for the current balance. In other words, change the state from transferring money to the flower shop to checking balance. At some point, the user will decide to return to paying the flowers.

The Oracle Digital Assistant platform enables you to work with these kinds of scenarios with built-in state management. You as a developer do not have to code and maintain the solution.

Enterprise Data Integration

When creating enterprise-grade Digital Assistants, it is somewhat obvious that these also need to have a way to access enterprise data (Figure 1-4). Oracle Digital Assistant facilitates connections to backend systems. It helps you extend backend systems such as Oracle HCM, Oracle ERP, and Oracle CX. However, you are not limited to Oracle products. You can also connect to third-party applications. All of this can be done, as can be expected from enterprise solutions, in a secure and scalable manner.

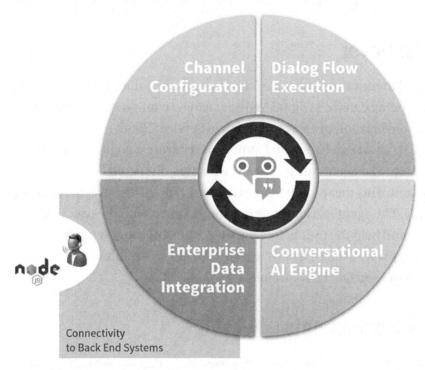

Figure 1-4. *Oracle Digital Assistant – enterprise data integration*

For the integration of enterprise data in Oracle Digital Assistant, you can use **custom components**. They provide your skill with generic functions like outputting text, or they can return information from a backend and perform custom logic. You as a developer can create custom components, developed in JavaScript and deployed onto a Node.js server. Custom components can be invoked during a dialog flow to retrieve information from backend systems or to perform transactions in backend systems. Of course, these actions will use APIs that you have to make available in your backend systems.

Another way to step out of the flow is by means of **QnA.** This is actually a very common use case for an assistant. Whenever the user types a phrase that matches a search term in the QnA, the matching questions and answers are displayed to the user. A developed skill can act as the interface to frequently asked questions (FAQs) or other knowledge base documents. Frequently asked questions are common questions that are looking for an answer: "What are your opening times?", "Can I bring kids?", "Do you have vegan pizzas?"

You can simply integrate a QnA service by importing sets of question and answer pairs from a simple spreadsheet. Both QnA and intents can be used in the same flow.

Agent Handover

Every now and then, a situation can occur where your chatbot users do need to speak to a real human being. Oracle's Digital Assistant can be configured in a way that it is able to hand over a conversation to an agent in Oracle Service Cloud.

The human agent can handle the inquiry as if it were a regular inbound communication. The agent can see exactly what was discussed between the user and the Digital Assistant. This means that there is no reason to ask the user to repeat themselves unnecessarily. The agent can continue the conversation with the user in a two-way conversation and help the user before passing control of the conversation back to the assistant. In this way, Oracle Digital Assistant provides a seamless collaboration between call center agents and Digital Assistants.

Webview Components

Typically, the conversation flow between a user and a digital assistant is really unstructured. For humans this is a very natural way of interacting. However, sometimes you might need a way to capture structured information, such as a data entry form. This is where you can use Oracle Digital Assistant Webview components. These **Webview components** are small self-contained modular applications that enable users to complete a simple task such as a task escalation or entering purchase details. They can be incorporated into the conversational flow, thus allowing the digital assistant to have both structured (webview) and unstructured interactions (the chat) with the user.

The user can switch from the natural language conversation with the bot to an app-like experience for structured data entry, instant validation, and rich media.

Within Oracle Digital Assistant, you find a web-based builder tool. With this tool, you can declaratively develop rich forms, including images, charts, maps, and signature capture. It is very easy to also add validations to ensure the accuracy of the data captured.

The Webview component can be invoked from a link inside the conversation. Once the data is entered, it can be passed back to the digital assistant, and the flow continues from where it was before entering the app.

Channel Configurator

The last component of the Digital Assistant's high-level architecture is the channel configurator (Figure 1-5). Digital Assistants are not applications that can be installed on a smartphone or tablet. Typically, they are made available over different **channels**.

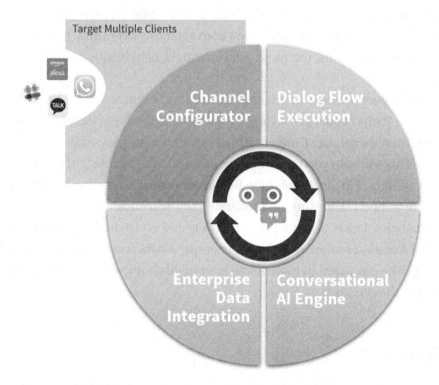

Figure 1-5. *Oracle Digital Assistant – channel configurator*

These channels, also referred to as messaging platforms or messaging apps, such as Facebook Messenger, Slack, and web pages, are the actual platforms where the user interacts with the digital assistant. Also text-only channels such as Twilio SMS and WeChat are supported. The channels are configured within Oracle Digital Assistant with their platform-specific configurations and allow access from the messaging platforms to the Digital Assistant and vice versa. One digital assistant or skill can have several channels configured for it so that it can run on different services simultaneously.

Actionable Insights

Another really valuable feature in Oracle Digital Assistant are the built-in analytics. These enable you to do in-depth analyses of conversations. It will show you how your Digital Assistant is used and where there are potential issues within the conversation. The analytics also help you to get some utilization information which can be used to improve the accuracy of the Digital Assistant, thus enabling you to create an even better experience for your end users. There are several instant reports available. The **Overview report** shows you a graph of all the conversation metrics. You can see how users have either abandoned or completed conversations over time. It will also help you to determine top intents, the most used channels, conversation duration, and error counts.

The **Intent report** provides intent-specific data and information for the execution metrics, and the **Paths report** gives you a visual representation of the conversation flow for an intent. Then there is the **Conversations report** that displays the actual transcript of the dialog. This will help you to view a conversation in the context of the dialog flow and the chat window. Finally, there is the **Retrainer report**. This report can be used to improve the developed skills through moderated self-learning.

All the analytics data is collected automatically and is also integrated with the test tool of Digital Assistant so that even during testing the data is collected and analyzed. The reports will be discussed in more detail later in this book where you will learn when and how to use these reports.

Summary

In this chapter, you learned about the history of chatbots and how Oracle Digital Assistant evolved from this. You were also provided with a high-level overview of the components involved in Oracle Digital Assistant and how these can help you to build enterprise-grade Digital Assistants. All of these components, and more, will be discussed in depth in the remaining chapters of this book.

PART II

Design and User Experience

CHAPTER 2

Designing a Conversational User Experience

Building digital assistants without a business case is a recipe for failure. You should always try to answer the key questions before you start like:

- What is the business pain we are trying to solve?

- What is the existing journey users take and how can it be improved?

- What is the appropriate channel to reach the users and how can it best be used?

- What is the conversation between the bot and user and the dialog option and how can the dialog be enhanced to use the channel's properties, media, and facilities to improve the user interactivity?

- What is the key vocabulary? Which are the entities exchanged in the dialog?

Answering these key questions is where the design phase has its value.

Why a Conversational User Experience?

Simply put, people are a lot better at talking with each other than using technology. That's because conversation is natural and innate. When speaking with a friend or a stranger, we can easily correct ourselves if we say something confusing or misspeak. Therefore people have been using conversation to drive sales and make customers happy since humans first began trading.

© Luc Bors, Ardhendu Samajdwer, Mascha van Oosterhout 2020
L. Bors et al., *Oracle Digital Assistant*, https://doi.org/10.1007/978-1-4842-5422-6_2

Conversational user experience (UX) makes the switch to a human form of communication as if a human is having a normal conversation with another person. Two popular ways of engaging with conversational UX are through vocalized speech and through typing text into a chatbot. Millennials appear to prefer to type over using the phone. And the rest of us are getting used to that too.

A conversational UX helps an enterprise to sell their products/services or helps customers more effectively. That is also the reason for this trend toward interacting with businesses through messaging and chat apps like Facebook Messenger, WhatsApp, Talk, and WeChat or through voice technology, like Amazon's Echo product, which interfaces with companies through voice commands. By means of conversational commerce, customers can chat with company representatives, ask questions, get customer support, get personalized recommendations, read reviews, and click to purchase all from within messaging apps. As more people become heavy mobile users, it's a great business idea to provide a more seamless experience of shopping online with one's mobile. With conversational commerce, the consumer engages in this interaction with a human representative, chatbot, or a mix of both.

Conversational commerce is powerful because it

- Makes the automated interactions between the brand and the consumer feel much more human. Conversation as an interface still is the most natural way for humans to interact with technology.

- Reduces the steps required to complete an action and the number of information sources consumers need to turn to.

- Enables two-way communication with the customer. It doesn't just tell them things but also learns from them, hears their questions, and builds a relationship. Delivering a personalized experience that caters to a user's active intent is critical to satisfying (and gratifying) usersandultimately earning their trust and loyalty.

- Is an extremely effective method to engage users/customers, collect information based on their needs, understand their active intent and goals, and then deliver an experience that has the greatest potential to satisfy these needs.

- Creates an interaction that is more personable and meaningful than for instance email. Not only does the language sound natural, like how a friend or family member would talk to a user, but it also enables the relationship to immediately strengthen by allowing the user to continue the conversation in one go, as you can see in the example illustrated in Figure 2-1.

Figure 2-1. *The user can continue the conversation with one click*

- Conversational commerce shortens the distance between prospect and purchase. It offers customers the attention they might get from a sales associate in the store. When ordering from a web site, they can read reviews to get a sense of whether a product will work for them, but using chat they can ask for help comparing their options – more like the advice they would receive in-store.

Beware of the fact that not all content is dialog-able. You do not want a chatbot to list a complete product catalogue or user manual. A conversational interface is a complementary interface.

When users require detailed instructions or tutorials, for example, video works better than text. This guarantees a usable and fixed information flow, something that is not guaranteed when using chat. A fixed flow of information works better in this case, not filling the chat dialog with a wall of text.

In essence, conversational UX is about making technology behave and interact more like we interact with one another.

Designing a Conversational User Experience

In order to design a unique conversational user experience, this chapter will disclose a method which guides you in eight steps to a result that satisfies all stakeholders – stakeholders like marketing managers, the development team, and last but not least the end user/customer, who all have their own specific requirements, wishes, and needs regarding the experience which in the end will be embodied in a chatbot. Figure 2-2 shows how all stakeholders are related.

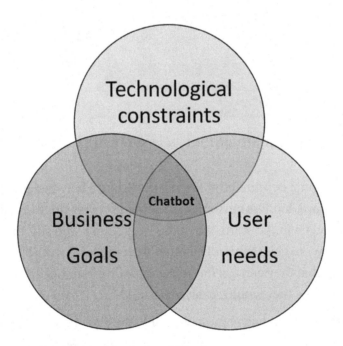

Figure 2-2. *The challenge is to find a good balance*

We've divided our process into eight steps. Figure 2-3 shows these steps. Then the sections that follow describe each one in sequence.

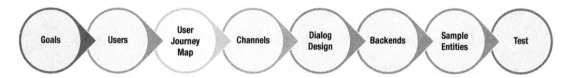

Figure 2-3. *The eight design steps*

The Goals

The first step in designing a great chatbot is asking yourself: "**Do we even need a chatbot?**"

The "why" of designing should always come before the "how."

—Will Fanguy, Inside Design

It is of great importance that the conversational user experience, and in the end your chatbot, serves a purpose. This is the foundation of everything. This should be something you'll do long before you even start to consider designing a conversational UX.

The moment you know what the purpose of your chatbot is, you have answered fundamental questions that need to be answered in order to move to the design process.

To find out which goals your client has in mind for your conversational experience, you can interview different stakeholders who know of the current business goals and have a vision for the future. Different stakeholders probably have different goals.

For example, a CEO might want to use the chatbot to

- Beat the competition

- Increase customer reach and customer awareness

- Reduce the number of service desk employees and therefore as a company be more cost-efficient

One the other hand, marketing managers want the chatbot to

- Seamlessly connect with the customer on Facebook Messenger

- Chat with the customer directly from the online store

- Recommend products based on purchase history and customer data that's been collected conversationally

- Touch base with the customer after they leave the store

- Ask the customer if they are ready to reorder an additional product

Finally, service desk managers might want to

- Let the customer know that their order has been received or shipped

- Thank the customer after their first purchase

- Confirm with the customer that their package arrived and ask for their opinion

- Do troubleshooting

- Answer questions

The main objective in design is to keep the goals simple, so do not get into all the constraints and requirements. Include enough so all stakeholders know what this project is trying to do, relative to everything else they themselves are trying to accomplish at the same time.

Prioritize the goals and keep them in mind while designing the conversational experience.

With every design decision, ask yourself: "Does this design get us where we're trying to go?"

The Users

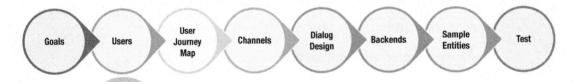

Figure 2-4.

As you saw earlier in the Venn diagram of Figure 2-2, the needs and wishes of the users play an important role when designing the conversational experience. That is the reason why you should know all about the users with whom you want your chatbot to communicate.

The first step for a user-centered design process is to do user research. Through this, for example, by observing and interviewing potential users, you can learn about

- Their goals and priorities

- Their behavior

- The daily habits

- Their education and skills

- Their knowledge regarding the domain of the enterprise

- Their use of tooling like a computer, a smartphone, Internet, social media, and other channels

Figure 2-5. *User research by interviewing and observing the target group*

Once you get all this interesting, useful rich data, the challenge is to communicate it across to the rest of the design/development team. An effective method we always use is creating one or more personas.

Creating Personas

A persona is a realistic character sketch representing a certain segment of the target group.

—Steve Mulder/Ziv Yaar, The user is always right

Personas actually are made-up people. The purpose of personas is to communicate all we know about the user and ensure that it is remembered and used throughout the project lifetime. For this to happen, the personas must be adopted by the design/ development team.

A persona has a name and is represented by a picture and a summary of some relevant characteristics. A quote, summarizing their needs, emotions, motivations, and goals into a single sentence, has the added advantage of giving the persona further personality. Figure 2-6 shows an example of how a persona may be described and illustrated.

Kelly Callomi

- 19 Years old
- Education: VMBO-T
- Living together with Michael
- Place of residence: Woerden
- Loves shopping, travelling and hanging out with friends

Helping customers in the world of Simpel©

She and Michael both **like to travel and to go out at night**, but these hobbies requires money. Kelly got her job at the **Simpel© call center** via her sister-in-law

QUALITIES

She is bright and learns by doing. She is patient, friendly and polite but also self-confident. She knows how to deal with hostile customers and how to work independently.

HOBBIES

She likes to go shopping with friends. She and Micheal often go out in the weekend. Once a year they go on holiday to a sunny destination where the beach and the bars are the place to be.

IDENTITY

Clothing and her scooter are very important to her. She has an IPhone which she is "addicted" to. She uses it more than she uses her PC at home. Mainly to keep in touch with her friends on Hyves and Facebook. She often checks her balance with the bank app.

EDUCATION & EXPERIENCE

She finished her VMBO-T. She knows how to perform office tasks and how to search the internet on a computer. She knows even better how to use her smart phone.

MOTIVATION

She likes working for the Simpel© call center. She is very motivated to help Simpel© customers: "a happy customer makes me happy". She also likes the possiblity for promotion within the department offered to her by Simpel©. However the moment she gets bored or she is offered a better job, she will resign.

SIMPEL© KNOWLEDGE

Everything she knows about the rules of Simpel©, she learned on the job. She herself is not a Simpel© customer. She considers herself working for Simpel© not for Aspider.

***Figure 2-6.** Example of a persona representing a call center employee*

There should be at least one persona to represent each major segment of your users. The word major is used because ideally you should be using three to five personas. Each of the personas should have their own personality and be memorable. If you have more than five personas, people often start to feel overwhelmed. And it seems that

When you design for your primary persona, you end up delighting your primary persona and satisfying your secondary persona(s). If you design for everyone, you delight no one. That is the recipe for a mediocre product.

—Alan Cooper

Benefits of Personas

Personas are beneficial because they can help to focus and target your thinking. Personas help you to think in less abstract terms about your audience and their needs. Benefits from carefully documenting user personas include the following:

- They bring focus on "the needs of real users..."

- They help designers/developers empathize with the users more easily.

- They are of use to many different stakeholders:

  - Business development (strategy development)

  - Marketing teams (seducing the users)

  - Design teams (design for the users)

- Personas encourage consensus within the team. It'll be clear when the personas have been adopted. Then the team starts referring to them on first names, "Kelly would never want to do that"... and others don't ask who Kelly is.

- Personas create efficiency.

- Personas can represent current but also future users. They help the team think more about the future use of the conversational experience.

- Personas motivate out-of-the-box thinking.

Everyone in the creative/development team now has an idea of how the users for whom they are creating the conversational experience look like. By putting up posters of the personas, after introducing the personas to the design/development team, they will be adopted more smoothly and grow on them during the development process.

Figure 2-7. *Life size personas on the wall will always be on the team's mind*

A persona enables the team to design the chatbot with knowledge about the user. What they develop, how it works, what you want to achieve, and last but not least the tone of voice can be based on what you know about the user.

The User Journey Map

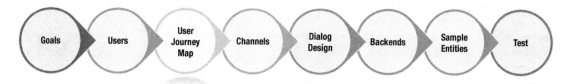

Figure 2-8.

Now that we have a clear view on the user(s) of our conversational experience, we need to figure out what journey he/she will make to achieve his/her goals using touchpoints/channels like the chatbot we will be creating. A well-known method to achieve this is defining a user/customer journey map.

The customer journey describes all the experiences that customers go through when interacting with your company's brand, product, service, web site, or application. The touchpoints[1] customers encounter are used to construct the journey and create an engaging story, detailing their interactions and accompanying emotions.

Instead of looking at just a part of a transaction or experience, the customer journey documents the full experience of being a customer. Conversational experiences can add value to every part of the customer journey, ranging from, for example, when a customer makes their first order to answering a product-related question instantly.

How to Create the User/Customer Journey Map?

It is crucial to identify the touchpoints where users interact with your product, service, web site, or application. Once the touchpoints have been identified, they can be connected in a sequence representing the way customers follow while interacting with the various touchpoints. Figure 2-9 shows an example of an overall customer journey map including all possible touchpoints and paths of users planning and making a short trip to Amsterdam via Schiphol airport using public transport.

---

[1] *A touchpoint is a contact point between the customer and the company or service provider. Touchpoints can be physical, virtual, or human. Examples of a touchpoint: a poster with an ad in an abri, a store, a service desk providing personal face-to-face contact, a vending machine, a web site, etc.*

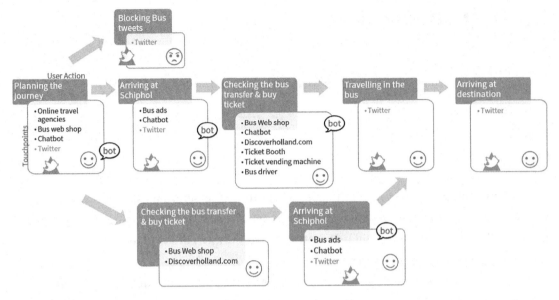

Figure 2-9. *Example of a journey map showing the different paths*

You analyze the activities customers perform and note down what they are doing during each activity in their journey.

Describe each of the relevant moments in the journey and try to identify the following:

- The context of use, where and when are they using the touchpoint.

- What do they want to achieve (their intent)?

- Which steps did they take to get there?

- What do they consider?

  - Do they have questions?

  - Are there uncertainties?

- What influences customers?

  - What motivates them?

  - Are there any discouraging obstacles? Think about anything that might cause the customer to give up while moving through the customer journey.

  - What emotions do they feel?

- What feedback do they expect?

This results in a visualization of a more detailed customer experience. Figure 2-10 provides an example of such a visualization.

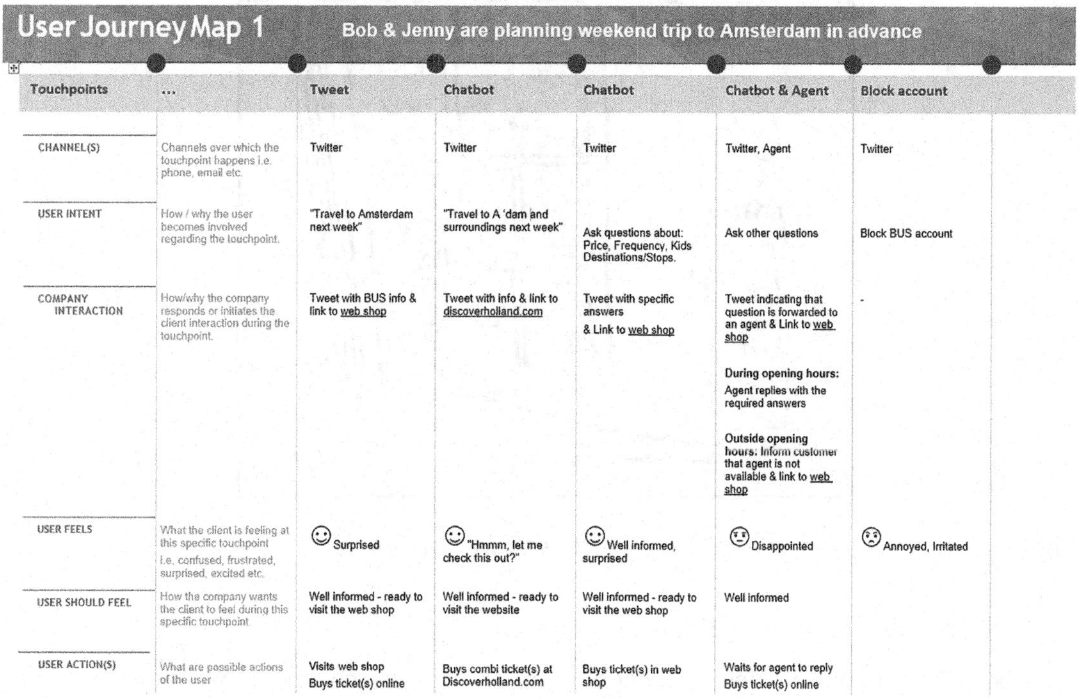

User Journey Map 1 — Bob & Jenny are planning weekend trip to Amsterdam in advance

Touchpoints	...	Tweet	Chatbot	Chatbot	Chatbot & Agent	Block account
CHANNEL(S)	Channels over which the touchpoint happens i.e. phone, email etc.	Twitter	Twitter	Twitter	Twitter, Agent	Twitter
USER INTENT	How / why the user becomes involved regarding the touchpoint.	"Travel to Amsterdam next week"	"Travel to A 'dam and surroundings next week"	Ask questions about: Price, Frequency, Kids Destinations/Stops.	Ask other questions	Block BUS account
COMPANY INTERACTION	How/why the company responds or initiates the client interaction during the touchpoint.	Tweet with BUS info & link to web shop	Tweet with info & link to discoverholland.com	Tweet with specific answers & Link to web shop	Tweet indicating that question is forwarded to an agent & Link to web shop During opening hours: Agent replies with the required answers Outside opening hours: Inform customer that agent is not available & link to web shop	-
USER FEELS	What the client is feeling at this specific touchpoint i.e. confused, frustrated, surprised, excited etc.	☺ Surprised	☺ "Hmmm, let me check this out?"	☺ Well informed, surprised	☹ Disappointed	☹ Annoyed, Irritated
USER SHOULD FEEL	How the company wants the client to feel during this specific touchpoint	Well informed - ready to visit the web shop	Well informed - ready to visit the website	Well informed - ready to visit the web shop	Well informed	
USER ACTION(S)	What are possible actions of the user	Visits web shop Buys ticket(s) online	Buys combi ticket(s) at Discoverholland.com	Buys ticket(s) in web shop	Waits for agent to reply Buys ticket(s) online	

Figure 2-10. *A detailed customer/user journey map*

Why Is a Customer Journey Map So Helpful?

Benefits from a customer journey map include the following:

- It provides you a high-level overview of the factors influencing the experience, constructed from a user's point of view.

- The map provides insight into the (future) use (the context, frequency of use) of the touchpoints.

- It enables the identification of both problem areas and opportunities for innovation. For example, could you improve the customer experience by proactively addressing questions your customers will have as they move through the stages?

- The different stages in the map can be used for further analysis and usage scenario[2] creation.

Figure 2-11. *The user plays the leading part in a scenario*

- It helps you prioritize.

- Throughout the design process, every major design decision will be motivated by the customer journey map.

What's Next?

After having worked out the user/customer journey map, you know which parts of the journey are crucial. Those are the ones that happen often or are of key importance to the user. Aim to help users to complete one of these journeys from start to finish and determine if they can be served by your chatbot design.

---

[2]*A story in which the user (persona) of the touchpoint plays the leading part. The story describes in more detail how the user interacts with the touchpoint and incorporates his thoughts, needs, and emotions.*

Channels

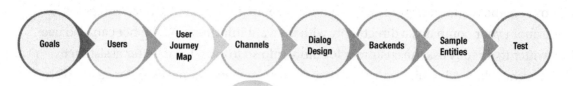

Figure 2-12.

The next step in your chatbot design is to determine what is the appropriate channel to reach the customers/users on their journey with your chatbot and how those channels can best be used.

The question of which channels should be used to reach the customer is two sided:

- What does the business think is the most effective channel to present the customer with the assistance of the chatbot? It might be one of the channels the enterprise already is using to reach their customers or even a totally new one which they see as an opportunity to reach even more customers and increase the sales of their products/services.

- In which channel does the customer expect, maybe even need, the assistance of the chatbot the most?

Facebook Messenger

Facebook Messenger chatbots are great for marketing and sales. They provide the means to engage prospects through conversations that allow companies to educate, evangelize, and entertain users around their products. The chatbot can allow also to follow up with users that abandon a cart in an ecommerce store and even to close the sale inside Messenger.

Many companies use a chatbot for prospecting and lead generation. In this case, they replace the typical "squeeze" page used to get their customers'/prospects' email addresses with personalized calls to action that include a "Send to Messenger" button. By doing this, users don't need to type anything. No email address or name. And of course, via FB Messenger, the company gets the users' real name (most of the time), the gender, and their system's language.

Facebook allows enterprises to integrate Messenger as customer chat on web sites. People can start a conversation in a web site chat window and continue it on Messenger.

Twitter

Your chatbot can send, customers who tweet about you, about a service you provide, or about your company, a direct reply with relevant information. The bot can also use Twitter to directly ask the customer for more info so to even be of better assistance.

Other channels which you might use for your conversational experience are Web, Slack, iOS, SMS, Google Home, Android, and Alexa.

Dialog Design

Figure 2-13.

The next step is to design your bot's dialog. It should engage users and move them toward their goals, and it should make sure that the companies' goals are achieved.

The dialog ideally consists of a mix of text and attachments like audio, video, images, and files. The dialog also can present menus and buttons which enable the user to give a quick reply. Figure 2-14 shows an example of such a dialog.

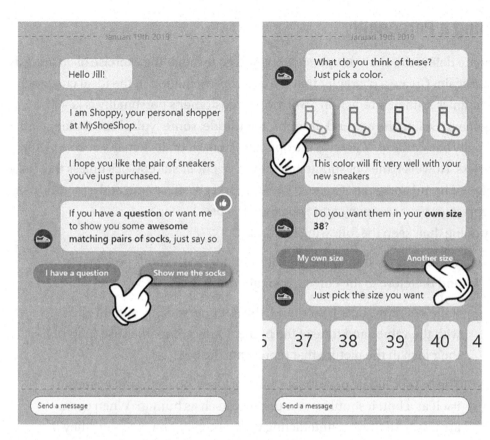

Figure 2-14. *Example of a dialog with options for quick reply*

The Chatbot's Personality and Tone of Voice

Every bot has a voice – which means every bot needs a personality. The personality of a chatbot is determined by its style, tone, and attitude.

Chatbots and voice assistants are to be used by humans. Conversational interfaces take care of better interactions between humans and computers. Through personality, we can personalize these conversations and make them more life-like, intimate, and representative of human interaction. Building a rich and detailed personality makes your chatbot more relatable, believable, and engaging to your users. It encourages customers to exhibit greater trust and stickiness.

Personality creates a better understanding of the bot and how it will communicate through choice of language, mood, tone, and style. If you do not create a personality for your chatbot, the users themselves will assign a personality to the bot anyway. Then you run the risk that their perception of your bot is not the one you want it to be associated with.

Building a Personality

In order to define the chatbot's personality, you go through the exercise of creating a persona again. Create a character that is a true representation of the ideal customer, which mimics human interactions. Mirroring a customer's personality is key to engagement. Define the age and describe an attitude, some typical vocabulary, and a personality background. You have the option to make it male, female, or completely genderless. However, the bot should not embody gender stereotypes unless that fits your companies' brand and goals. Create a persona, and give it a name and a picture/icon that represents your company and appeals to the user.

By doing that, keep the following in mind:

- The customer should emotionally connect to the bot's personality and therefore to the conversational experience.

- If the bot represents your brand, the bot's personality should align with the values and tone of the brand. Many companies use (part of) their logo in the picture that represents the bot.

- The bot should be personable, but not a person. It should express itself as a bot; it should never disguise itself as human. When users are unsure if they are interacting with a human or a machine, they'll feel misled, lose trust, and have a poor experience. Customers should know they are dealing with a bot. This, for example, can be achieved by letting the bot introduce itself as a bot.

"Hi! I'm Cuby. I am a chatbot. How can I help you?"

The Tone of Voice

Now that you defined the personality of the chatbot, you need to understand how the bot should "speak." You can figure that out by writing down key values of the bot. You define the chatbot's attitude, style, and tone which should reflect its personality. Then you decide what vocabulary should be used.

You then can bring together the personality you designed and the voice you envisioned, to write your bot messages.

For example, the bot as a cool dude:

"Hey, Harold, wassup? I see you're checking out this super chill hiking holiday. How can I help you out?"

Or a serious and professional bot:

"Hi Amy, I see you're editing your timesheet. Can I be of help?"

The Conversation Flow

The conversation flow can be visualized in a flowchart. To create the conversation, the customer/user journey map is of great use. In the map, you can find the user intent and how the bot (which is representing the company) is supposed to react. The user intent can be deducted from what the user is doing at the time or by a question from the bot like a conversation starter:

"Hi! I'm Cuby. I am a chatbot. Can I help you with your booking?"

Designing a Motivating Dialog

It is essential that the chatbot causes joy and motivates users to act. The bot causes joy when it fulfills the users' needs and wants. Motivation enables a certain behavior. Users can be motivated by one or more of the following factors:

Information

The information given by the bot should be relevant to the user, easy to understand, and trusted and enable relevant choices. It should also be easy for users to know what the bot can or cannot do, in order to set the correct user expectations. The bot should clearly communicate its core functionality. If user intent is misunderstood or not achievable by the system, be honest and let them know they need to try a different approach.

"Did you know I can help you with …?"

"I'm sorry, I can only help you with questions regarding flights to the USA"

"I'm sorry, I do not know what you mean, can you please rephrase the question?"

Goal Setting and Commitment

People set their goals through a comparison of the present and the desirable future. As a designer of the bot's dialog, make sure you know the customers' goals and guide them toward those goals.

The moment users are committed to their goals, it is more likely that they will pursue that behavior and stay focused.

Train the bot to engage the user on personal topics by creating a more engaging and more personal chit-chat dialog. By doing that, the bot will learn to predict information about future users too.

"Hey Jenny, how are you doing today?"

Incentives

Incentives or rewards motivate users to a certain positive behavior. Incentives and rewards do not always have to be economical; status or convenience may also have important effects on how users respond to the bot.

"If you tell me when it's your birthday, I can send you a birthday card."

Autonomy

Being the cause of your own actions and doing things in your own way is what users motivates. Therefore, the bot should steer users but at the same time make them feel in control. You can, for example, achieve this by offering the user a way out at any time. Figure 2-15 provides an example. Notice the "I prefer to do it myself" button near the bottom of the right-most panel.

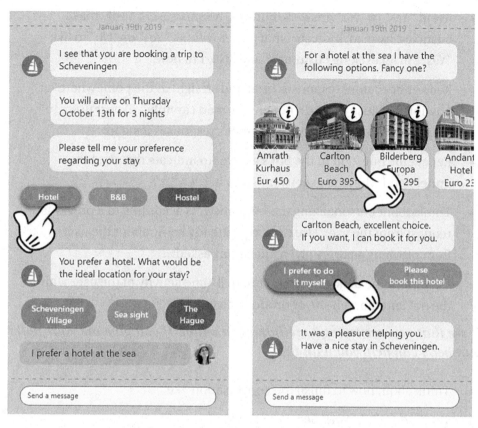

Figure 2-15. *Example of a dialog that makes the user feel in control*

Feedback

One of the most important factors to motivate positive behavior toward the bot is feedback. Feedback presented by the bot is needed to communicate some of the previously mentioned motivation factors. For example, goal setting requires feedback to communicate performance toward a goal.

"You've selected the men's department."

Other issues to keep in mind while designing the dialog are as follows:

- Keep the conversation topics close to the purpose served by the chatbot. If the user is drifting off, try to get him/her back to the main topic.

- Avoid "Yes" or "No" answers/buttons, because users easily make mistakes. Instead, indicate what a user will say "Yes" or "No" to like "Yes, I'm an existing customer" and "No, I'm new."

- Keep your chatbot responses brief and straight to the point wherever possible because users do not always read carefully. Expect users to frequently skim through your chatbot's responses instead.

- Repeat the question in the bot's answer to indicate that the bot understood what the user wants from it.

- Think about personalizing the bot's answers according to a previous request from the user. Use what you already know about the user from previous answers given/questions asked.

- Customize the dialog based on the real-time activity of the user with the bot and the web site or product it supports.

"I see that you are trying to reimburse a claim. If you give me your policy number, I can help you with that."

- Think about how to finish the conversation like

"Always glad to help you! See you soon."

Backends

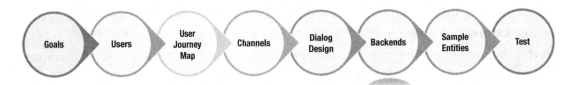

Figure 2-16.

When you design, your digital assistant must cater for the fact that it will not have all the knowledge built in that is needed to handle the entire flow. For example, authentication will not be implemented in the digital assistant itself but will be provided by a backend system. Data such as customer data, employee data, and stock information will not be part of the digital assistant itself. In an enterprise context, there might be a need for a human resource bot/Skill that can execute HR-related tasks like employee profile update and leave balance or a recruitment bot for recruitment-related activities. Information to execute such tasks is typically available in your backend applications. Your digital assistant should be able to connect to these systems and bring information from these systems into the conversation.

Let's take the example of the Digital Assistant from Figure 2-17.

Figure 2-17. *The user wants to book the selected hotel*

In this example, the user chooses to book a hotel. For the DA to do this, the DA needs to communicate with the backend booking system. When the DA invokes the backend system, for the user, the hotel will be booked as required. Typically, the DA sends a request, containing the booking details, to the booking system which then takes care of the actual booking. The user has no clue that the DA calls out to a backend system. It should be completely transparent and seamlessly integrate with the UX flow.

In your design, you should recognize all parts of the conversation where backend integration is required. You also need to provide the information that is needed to properly create this integration. Think, in the context of the booking example, of check-in and check-out dates, hotel name, and last but not the least the username that is booking the hotel. Providing this information enables developers to properly create the APIs and components that implement these points of integration.

Sample Entities

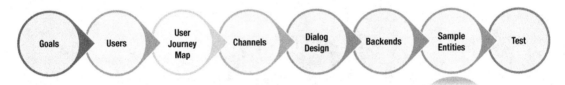

Figure 2-18.

As described in the journey map, with each touchpoint, the user has one or more intents, for example, "Travel from Schiphol to Amsterdam." An *intent* is the user's intention.

An *entity* modifies an intent. For example, in the intent "Travel from Schiphol to Amsterdam," the entities are "Travel," "Schiphol," and "Amsterdam."

Anything the user literarily says is called an utterance. For example, if a user types or tweets "I just arrived at Schiphol for a city trip to Amsterdam," the entire sentence is *the utterance*.

When creating a flowchart for the conversation flow, except for the happy flow, you will have to formulate different utterances, because there are so many ways for a user to express himself/herself.

"Waiting at Schiphol. Looking forward to my city trip to Amsterdam."

"Just arrived at Schiphol for a weekend trip with my whole family."

"Our weekend trip to Amsterdam starts here at Schiphol airport."

A combination of entities forms the basis for these different utterances. The chatbot in the end must recognize the user's question/remark – regardless of how it is written – to be able to come up with the right answer.

It can be helpful to share the entities with project stakeholders. Using the word cloud approach shown in Figure 2-19 is a quick way to provide an at-a-glance view of the supported entities that also shows their relative importance or at least their relative usage.

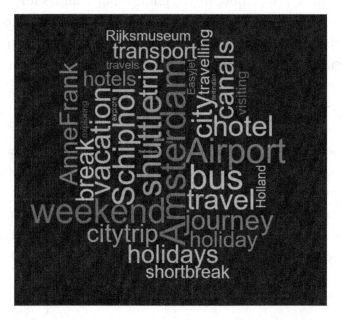

Figure 2-19. *A word cloud is a way to present all different entities to the stakeholders*

The Usability Test

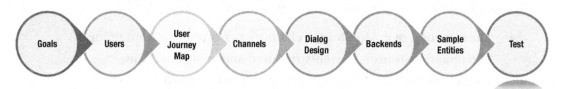

Figure 2-20.

This is where all the hard work pays off. While this is the eighth step that's outlined here, it doesn't necessarily have to be the last. As a matter of fact, most of the time, the information and feedback you gather in the test stage lead to redefining the problem, finding more sample entities, and better empathizing with your users. Keep asking yourself the question: Are we building the right thing?

Using Test Scenarios

You created a flowchart for the conversational flow. Each path through the flowchart from top to bottom is a test case and can be used to test the happy path only.

What the flowcharts don't show are the endless possibilities for a user to express an intent and the variance in conversation responses by the bot itself. However, a well-trained chatbot does understand the different user expressions (utterances), and a well-designed chatbot provides some variance in conversation responses.

Exploratory Testing by a Lot of Users

It is important to select the crowd wisely depending on the project state. The most important selection criteria are language and dialect. For this test, there is not a need to use any requirements or test cases; just let the crowd do their work.

Blind testing can also be a good way to test. Can users who are in the target audience but who have not been trained in using your chatbot succeed in using it the first time they are confronted with it? That's a good question to ask, and user testing will provide the answer.

A/B Testing

A/B testing is defined as comparing two versions of the conversational UX by presenting both versions to similar visitors at the same time. The goal is to see which one performs better.

By investing in an A/B test, you will be able to measure the:

- The retention rate: Monitor the chatbot's performance in attracting and retaining its users. A well-structured and engaging UX in messenger chatbots may boast retention rates up to 70%, unlike widely used email marketing campaigns with up to 40%.

- The drop-offs: The place in the dialog flow where the user left an intended main flow and went down a path different than the intended flow. This means that there might be some content or elements of the UX that visitors are not reaching because they get bored or frustrated before reaching them.

- The main conversions: Test which conversational UX is best leading users to what is most valuable to your business, for example, clicking a particular "Call to Action," starting a live chat with an agent, completing the checkout process, and so on.

Some of the chatbots' building platforms offer ready-made solutions for A/B testing of the chatbots. You can use them for this purpose.

Every experiment that is set up for A/B testing may include two plus variations of the same feature or UX element that requires validation. You can even do that when your bot is live. Try different utterances and measure changes in user behavior.

CHAPTER 3

Use Case

Introduction

To make sense of the technical implementation of Oracle Digital Assistant, in this chapter, you will be introduced to Travvy, the Digital Assistant/chatbot for Extreme Hiking Holidays, a travel agency – made up by the authors – specialized in hiking holidays for adventurous people.

Figure 3-1. *Extreme Hiking Holidays company web site*

Extreme Hiking Holidays organizes high hiking scale tours in Switzerland/France, Canada, and the USA. They offer a complete travel package including flights, a guided hiking tour, and accommodation.

Using a step-by-step approach, you will get acquainted with Travvy so you know all the ins and outs in order to understand all the steps of the creation of the conversational experience and the technical implementation.

© Luc Bors, Ardhendu Samajdwer, Mascha van Oosterhout 2020
L. Bors et al., *Oracle Digital Assistant*, https://doi.org/10.1007/978-1-4842-5422-6_3

You will read about the **"why" of Travvy,** and you will meet **the persona** which represents the traveler/user who will be communicating with Travvy. The **user's journey** is worked out explaining Travvy's role in it. You will see how a **channel** for the Digital Assistant is selected, and finally you will find the **dialog/conversation flow** revealing the user's intents. In the last part of this chapter, the **entities and utterances** used for the chatbot are presented.

The Goal

Who Are the Stakeholders?

The CEO and marketing manager of Extreme Hiking Holidays are the decision-makers.

Extreme Hiking Holidays is a small travel agency. Three employees take care of the reservations, and six adventurous guides are on site accompanying the travelers. The agency has a large market share in an emerging niche market.

Why a Chatbot?

The marketing manager knows that their customers have a lot of questions regarding the risks of such a holiday, the level of difficulty, and the insurance conditions. The reservations department currently is responsible for answering all these questions by mail or on the phone. Questions which often have the same answers.

A chatbot helps reducing costs, because a person answering a phone call or email often costs more than a chatbot answer, which only costs less than a cent. That also is the reason why a chatbot should rather not refer to a phone number. Better provide users with a link, so that they themselves can find the requested information.

That is the reason why they are thinking of developing a chatbot, to be able to offer their customers a better service, 24/7, advise/help them, and persuade them to book a holiday with them.

The advantage of a good chatbot is that the user can ask anything. This means that there is no need to search/browse a web site. Users can just ask what they are looking for. The digital assistants can even display a requested form so that the user can complete it immediately. **Smart digital assistants become smart search engines**.

The Customer: A Persona

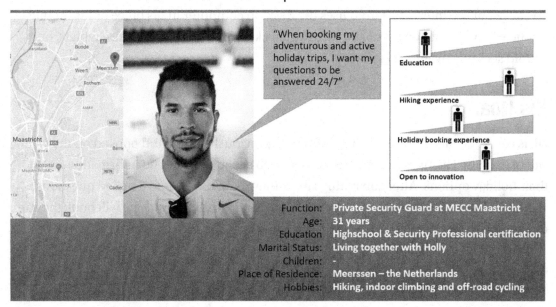

HAROLD HIKAWAY | *FOND OF ACTIVE HOLIDAYS*

"When booking my adventurous and active holiday trips, I want my questions to be answered 24/7"

Education

Hiking experience

Holiday booking experience

Open to innovation

Function:	Private Security Guard at MECC Maastricht
Age:	31 years
Education	Highschool & Security Professional certification
Marital Status:	Living together with Holly
Children:	-
Place of Residence:	Meerssen – the Netherlands
Hobbies:	Hiking, indoor climbing and off-road cycling

Figure 3-2. *Persona Harold Hikaway (Photo by Danijela Froki on Unsplash)*

Typify Harold

In his spare time, Harold is very active. Every Saturday morning, he goes on a long cycling trip with a team of friends. He and Holly, his girlfriend, go climbing indoors once a week, and during their holidays, they like to do that in the Dolomites (challenging) and in Czechia (cheap). They both like to explore new hiking trails. Harold is to the point ("cut the crap") and expects others to be direct and honest too. He dislikes every minute he cannot spent in an active way.

His Working "Day"

Harold often works in night shifts. His main task then is securing the Maastricht Exhibition & Conference Centre (MECC) building in Maastricht. During those nights, he does his rounds, but in between he often has some time to kill, and that is when he is searching web sites to find information about adventurous trips and nice holiday destinations. He once booked a trip to France on the Extreme Hiking Holidays web site. He and Holly thought that was an amazing adventure. To be able to book that hiking trip, he at that time created a profile on their web site.

His Tooling

To check out the Internet, Harold uses the desktop computer in his office or his mobile phone when he is doing his onsite security checks. Some of his friends use tablets, but Harold sees no advantage in that.

His Goal

He is very much into active holidays which he previously found and booked on the web site of Extreme Hiking Holidays. He uses the web site to get detailed information about their holiday options. After consulting Holly, they often have some detailed questions. If these can be answered quickly – preferably during his night shift – he will book the trip on the web site in one go.

The User Journey Map

What journey will Harold make to achieve his goals using touchpoints/channels like our chatbot?

The Touchpoints

- Advertisements on Internet/TV at home on the couch
- Facebook at home or at work
- Extreme Hiking Holidays web site at home or at work
- Chatbot

Harold's Intentions

- Check out some hiking holiday options.

- Find and select a hiking holiday and ask related questions.

- Book and pay for the hiking trip.

His Motivations/Needs

- Harold is triggered by the attractive holiday options Extreme Hiking Holidays has to offer.

- He notices that he does not have to enter his credentials.

- He gets clear, fast, and personal answers to his questions, 24/7.

- He likes to be guided, but he doesn't want to lose **autonomy**. He wants to feel in control.

Emotions to Be Triggered

- He is **pleasantly surprised** that the chatbot remembers his preferences.

- The tone of voice of the chatbot should

  - Be **appealing** to him

  - **Amuse** him

  - **Stimulate** him to act

- He should be **satisfied** with and **confident** about the result.

The following Figure 3-3 shows the overall journey Harold makes while searching for and committing to a hiking holiday.

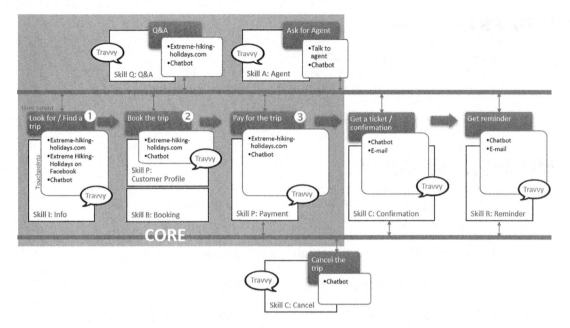

Figure 3-3. *Overall journey and touchpoints*

For the chatbot we will be creating in this use case, the focus will be on the first three parts of the journey in the core area and the two parts at the top.

For four core parts, you will find the total detailed customer journey in the next figures.

Because the chatbot only is one touchpoint of many the user engages with to find and book a trip, it is essential to work out the whole customer journey to make sure that no interaction is missing.

In this book, we are focusing on the user's interaction with only one touchpoint, the chatbot. Therefore, in the following pictures, the chatbot part of the journey is highlighted.

As you can see in Figure 3-3, the first thing Harold does is checking out the hiking holiday options. By doing so, he encounters several touchpoints like advertisements on TV and the online information from Extreme Hiking Holidays, and he will meet their chatbot.

Figure 3-4 presents a worked-out user journey describing what Harold's main intent is, his actions, and the company activities to support him.

Harold is checking out some hiking holiday options				
Touchpoints *Combination of channel, device and user task/intent*	**Advertisements** triggering user interest	**Check out Extreme Hiking Holidays online** browsing for holiday options	**Communicate with Website/Facebook chatbot** digging into Extreme Hiking Holiday details	**Contact Extreme Hiking Holidays online** for more specific info
Channels *Channels over which the touchpoint happens i.e. phone, email etc.*	Watching TV / surfing on the internet	Internet, social media	Chatbot on website/Facebook	Agent
User intent *How / why the user becomes involved regarding the touchpoint*	Being curious because of an appealing add from Extreme Hiking Holidays	Globally checking out the travel options and reviews	Be guided according to his travel wishes. Ask questions e.g. about availability.	Ask other questions
Company interaction *How/why the company responds or initiates the client interaction during the touchpoint*	Promote hiking trips and refer to website & Facebook page	Offer appealing and easy-to-find online content	Chatbot guiding and advising Harold to find the best trip and answering Harold's questions	1. ☺Agent is available 2. ☺Agent is NOT available, chatbot help is enough 3. ☹Agent is not available, chatbot help is not enough
User should feel *How the company wants the client to feel during this specific touchpoint*	☺Well informed - ready to visit the website/Facebook page	☺Well informed – be inspired by the information given	☺Well informed, guided, happy, ready to select and book a trip	☺Well informed, guided, ready to select and book a trip
User actions *What are possible actions of the user*	Visit website or Facebook page	Check out the holiday options	Indicate personal preferences and find & select a trip	Waits for agent to reply, talk/chat with the agent.

Figure 3-4. ❶ *Harold finds a challenging holiday*

In the journey map, you can also describe how you think your user should feel while encountering the touchpoint, what channel will be used for the encounter, and even the context of use. Because the presumed context of use for Harold during this journey mainly will be his desktop PC at work or at home, the context is not described in these journey maps.

After Harold found the desired trip – and of course checked it with Holly – he wants to make reservations and book it. See Figure 3-5.

Harold is booking the trip			
Touchpoints	*Combination of channel, device and user task/intent*	**Extreme Hiking Holidays online** for booking the trip	**Contact Chatbot** for booking assistance
Channels	*Channels over which the touchpoint happens i.e. phone, email etc.*	Website/Facebook	Chatbot on Website/Facebook
User intent	*How / why the user becomes involved regarding the touchpoint*	Book the selected hiking trip	Get assistance while booking the selected hiking trip.
Company interaction	*How/why the company responds or initiates the client interaction during the touchpoint.*	Make sure that the booking is completed with the correct booking details	Help the customer to complete the right data and book the trip in an efficient way.
User should feel	*How the company wants the client to feel during this specific touchpoint.*	☺ Sure, that he made the right choice. That the booking did not take much time	☺ Surprised that the chatbot knows certain things about him. ☺ Happy with the guidance. ☺ Well informed, confident and relieved that he doesn't need to worry about this booking.
User actions	*What are possible actions of the user*	Indicate which trip he wants to book.	Follow the guidance of the chatbot and book the trip.

Figure 3-5. ❷ *Harold books the selected hiking holiday*

A booking is not complete without payment. Harold can pay for his booking online via the web site, being assisted by the chatbot and in close contact to his bank. See Figure 3-6.

Harold is paying for the trip				
Touchpoints	*Combination of channel, device and user task/intent*	**Extreme Hiking Holidays online** for paying the trip	**Contact online chatbot** for payment assistance	**Contact Bank** online to withdraw the money
Channels	*Channels over which the touchpoint happens i.e. phone, email etc.*	Internet, social media	Chatbot on website/Facebook	Mobile app / Website of the bank
User intent	*How / why the user becomes involved regarding the touchpoint*	Pay for the booked hiking trip	Pay for the booked hiking trip	Withdraw money to pay for the trip
Company interaction	*How/why the company responds or initiates the client interaction during the touchpoint.*	Offer an appealing and easy-to-find way to pay for the booking – call to action button "Pay now"	Chatbot guiding the user through the payment process	Make save and secure payment possible
User should feel	*How the company wants the client to feel during this specific touchpoint.*	☺ Sure, that he did pay in a safe way. That the payment did not take much time	☺ Surprised that paying is so simple. ☺ Happy with the guidance. ☺ Confident and relieved that he paid for the trip in a save way.	☺ Confident about a safe payment
User actions	*What are possible actions of the user*	Provide personal payment details and pay	Follow the guidance of the chatbot check his balance and pay for the trip.	Identify himself, check his balance

Figure 3-6. ❸ *Harold pays for the booked holiday*

Of course, after all these actions, Harold will need feedback about his booking and about the payment being successful. He will receive that online on his personal page, "My Extreme Hiking Trips," but also the chatbot can provide this information and his tickets/vouchers. See Figure 3-7.

Harold is checking the confirmation of his booking				
Touchpoints	*Combination of channel, device and user task/intent*	**Extreme Hiking Holidays Online** check status of booking	**Online chatbot** proactively sharing booking status	**Mailbox** for receiving travel documents
Channels	*Channels over which the touchpoint happens i.e. phone, email etc.*	Internet, social media	Website/Facebook	Mailbox (mobile / PC)
User intent	*How / why the user becomes involved regarding the touchpoint*	See the status of his booking	Get feedback about the status of his booking Receive tickets/vouchers and store them	Get feedback about the status of his booking Receive tickets/vouchers
Company interaction	*How/why the company responds or initiates the client interaction during the touchpoint.*	Inform the user where to find confirmation of the booking and how to get the tickets/vouchers	Chatbot informs the user about the status of his booking, providing tickets/vouchers, helping him to store/print them	Send an email to confirm the booking of the trip + an attachment with the tickets/voucher.
User should feel	*How the company wants the client to feel during this specific touchpoint.*	☺ Confident	☺ Happy with how easy this was. ☺ Confident and looking forward to his hiking holiday	☺ Confident. He now can look forward to the moment he will go on holiday.
User actions	*What are possible actions of the user*	Check out the booking details and print out the tickets/vouchers or store them on his mobile	Follow the guidance of the chatbot. Store/print the travel documents	Store/print the travel documents

Figure 3-7. ❹ *Harold gets the confirmation of his booking*

The Channel

The touchpoint, mostly used by Harold to communicate with Extreme Hiking Holidays, is their Facebook page and their web site. That is the reason why for this use case the chatbot will interact with the user on those two channels: Facebook and the web site. Of course, Harold will be able to communicate with the chatbot on his computer and also on his mobile.

The Dialog

The Bot's Personality

The users of our chatbot are young, active, and sporty like Harold. They are hardly ever bored. When not working, they are hiking and doing (team) sports. They often are among friends. Then they talk in detail about their sporty activities, hiking trips, and holidays.

The marketing communication activities from the "Extreme Hiking Holidays" company fit with that image. Therefore, the chatbot's **tone of voice** should be as if it is one of Harold's cycling team members. The bot's **name** should fit to the company's image. Because the bot is going to offer a travel-related service for the "Extreme Hiking Holidays" company and because it should not be too serious, it will be called Travvy. Its "**portrait**"/avatar is derived from the company logo. See Figure 3-8.

Figure 3-8. *Options for Travvy's "portrait"*

Travvy is **genderless**, although testing the chatbot with users should confirm if they agree with this statement.

The Dialog Flows

Harold's main intents form the basis for the dialog flow. As you can see in Figure 3-3, each phase in the overall journey is represented by an intent which requires a certain skill of the chatbot. In the first phase, the "Info" skill of Travvy will assist Harold with his intent to find the right trip for him and Holly and indicate their preferences to Travvy.

Figure 3-9 represents the dialogue flow for Harold to find a trip.

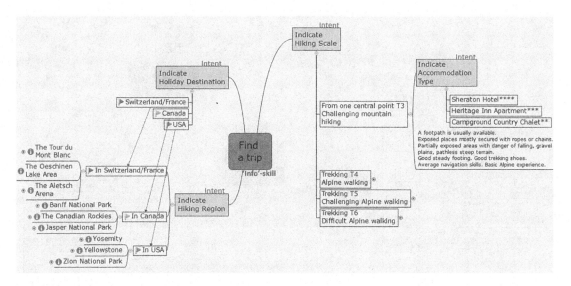

Figure 3-9. *Dialog flow to find a trip*

In this first phase of his journey, Travvy will ask Harold to indicate his holiday destination (country and region) and also the level of difficulty for the hiking trip that he has in mind. If he decides to take level 3 hikes from one central point, he can indicate what kind of accommodation he prefers.

After Harold found and selected the desired holiday, Travvy can assist him booking that tour. Therefore, Travvy will incorporate two skills: a "Customer Profile" skill and a "Booking" skill.

Figure 3-10 represents the dialogue flow for Harold to book his trip.

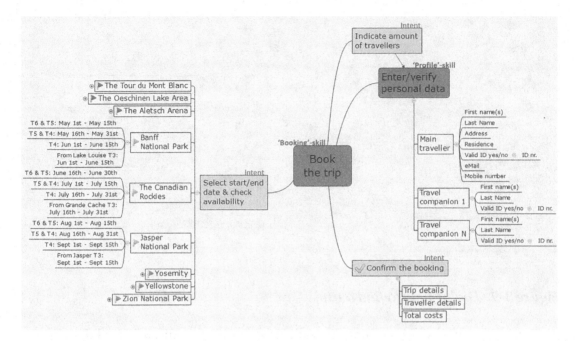

Figure 3-10. *Dialog flow to book the trip*

Harold will be asked to select a travel period and to verify/enter all the personal details of himself and his travel companion.

Now that Harold booked and confirmed his trip, Travvy will ask him if he likes to be assisted with the payment. To be able to do that, Travvy contains an instant app which provides the "Payment" skill.

Figure 3-11 represents the dialogue flow for Harold to pay for his trip.

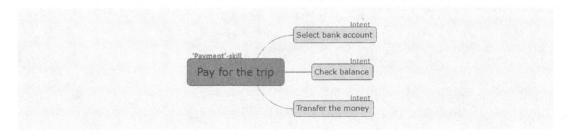

Figure 3-11. *Dialog flow to pay for the trip*

Travvy will ask Harold to select his bank account, to check if he has enough balance, and then to transfer the money.

In order to find out if the payment went well, Harold requires confirmation. Of course, he also wants to receive the trip-related tickets and vouchers. Therefore, Travvy contains a customer component which provides the "Confirmation" skill.

Figure 3-12 represents the dialogue flow for the confirmation the digital assistant will provide.

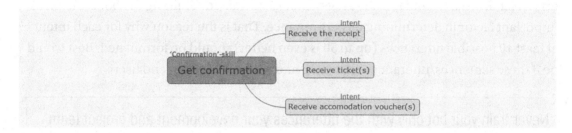

Figure 3-12. *Dialog flow to get confirmation*

Travvy will give him a receipt and can help him save/print out his tickets and vouchers.

Backends

During the conversation, Travvy requires Harold to enter personal data and payment information. Even though this data is needed for the conversation, it will not be stored inside the digital assistant itself. Typically such information will only be available from and stored in backends. Backends are the information systems that contain the data that is needed to support the dialog. The chatbot must be able to connect to these backend systems so it can include information from these systems in the conversation.

Backends will have to provide REST interfaces that can be used to allow integration with the digital assistant. Within the digital assistant, custom components need to be coded for the actual integration with backend systems. One other way to integrate with backend systems is to use forms for data entry. Travvy uses a form to allow Harold to enter his personal data directly into the backend.

Entities and Utterances

What Users Say/Write

In the preceding paragraphs, we defined the user's journey and formulated his intents. Harold can formulate his intents in so many ways (utterances). To make sure that Travvy understands what Harold means, Travvy should be trained. Training the bot is the most important factor in determining its performance. That is the reason why for each intent at least 10 possible utterances (up to 50 is even better) should be formulated. Best would be if these sentences/utterances are varied and really come from end users.

Never train your bot only with the utterances your development and project team come up with: they know the technical slang too well to accurately represent the real users like Harold.

Figure 3-13 shows different utterances for one of Harolds intents.

Indicate Holiday Destination					
Harold says	Harold's choice of destination		Travvy's feedback	Travvy's answer	Options
I'm thinking of going to [Holiday Destination] this year	Switzerland	>	Top! You go for Europe.	Which region in Switzerland/France is your favorite?	•The Aletsch Arena •The Tour du Mont Blanc •The Oeschinen Lake Area
[Holiday Destination] it will be	Canada	>	Great, Canada is well known for its forests, lakes and mountains.	Which region in Canada is your favorite?	•Banff National Park •The Canadian Rockies •Jasper National Park
[Holiday Destination] is favorite to me and my partner	USA	>	All our USA trips are in famous national parks.	Which region in the USA do you prefer?	•Yosemity •Yellowstone •Zion National Park
I want to go to [Holiday Destination]					
I want to visit [Holiday Destination]					
Your best hiking trips are in [Holiday Destination]					
[Holiday Destination] would be my first choice					
In summer, [Holiday Destination] has the best weather for an extreme hiking holiday					
Don't want to travel that far, so [Holiday Destination] it will be					
I think [Holiday Destination] offers the most challenging hikes					

Figure 3-13. *Utterances for the intent "Indicate Holiday Destination"*

What Travvy Says/Answers

Keep in mind a chatbot is a conversational interface. A conversation should be interactive. Travvy should never reply with large blocks of texts (more than 60 characters is too long). Therefore, the reply should be short, and images, buttons, lists, and other UX components should be used to make the conversation lively and rewarding and keep a user like Harold engaged.

Figure 3-14 shows how the interaction can be presented to the user.

Figure 3-14. *Using buttons and images to make the interaction easier for the user and more accurate for the bot*

Next to this way of presenting the dialog, you must make sure that the entire flow can be done using natural language. An example of a natural conversation can be found in Figure 3-15.

Indicate Holiday Destination					
Harold says	Harold's choice of destination		Travvy's feedback	Travvy's answer	Options
I'm thinking of going to [Holiday Destination] this year	Switzerland	>	Top! You go for Europe.	Which region in Switzerland/France is your favorite?	•The Aletsch Arena •The Tour du Mont Blanc •The Oeschinen Lake Area
[Holiday Destination] it will be	Canada	>	Great, Canada is well known for its forests, lakes and mountains.	Which region in Canada is your favorite?	•Banff National Park •The Canadian Rockies •Jasper National Park •Yosemity
[Holiday Destination] is favorite to me and my partner	USA	>	All our USA trips are in famous national parks.	Which region in the USA do you prefer?	•Yellowstone •Zion National Park
I want to go to [Holiday Destination]					
I want to visit [Holiday Destination]					
Your best hiking trips are in [Holiday Destination]					
[Holiday Destination] would be my first choice					
In summer, [Holiday Destination] has the best weather for an extreme hiking holiday					
Don't want to travel that far, so [Holiday Destination] it will be					
I think [Holiday Destination] offers the most challenging hikes					

Figure 3-15. *Travvy's feedback and answers for the intent "Indicate Holiday Destination"*

It is very important that Travvy gives Harold **clear feedback** regarding Harold's questions/answers. This feedback will make sure that Harold is **confident** about the fact that Travvy understood him well.

All other possible user utterances and bot replies defining the dialog can be found in the related Excel sheet.

By taking all these tips into account, the conversational experience for the scenario that Travvy helps Harold select a holiday destination and books the journey for him might look as visualized in Figure 3-16.

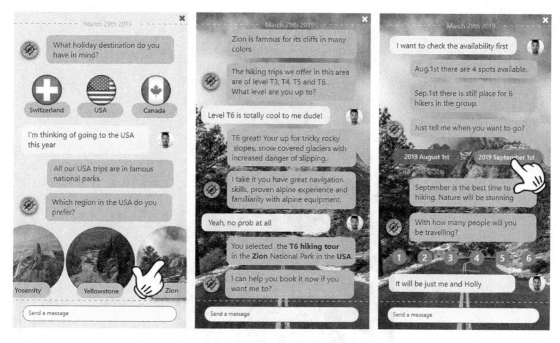

Figure 3-16. *Harold selects a holiday destination and a travel period*

Harold can use the buttons, but he can also type/enter his answer. If your chatbot is voice driven, Harold can even "speak" to Travvy.

The moment Harold selects a trip, Travvy can help him do the booking by asking Harold to verify/edit the personal data of himself and his travel party. See Figure 3-17.

Figure 3-17. *Travvy asks Harold to verify/edit all personal data needed for the booking*

Travvy uses the "Customer Profile" skill to prefill some of the data already known. To display and collect that data, Harold will see a different screen. So that he also knows that this is a side-step from the chatbot dialog. In this screen, he is not chatting with Travvy but seriously filling in personal data.

This additional screen might look like as presented in Figure 3-18.

Figure 3-18. *Harold verifies his own profile and edits the personal data of his travel party*

If Harold does not want to edit the data now, he can cancel this screen to go back to the conversation.

The moment Harold entered and verified the data, he can close the personal data screen by clicking OK and continue with the conversation as for example visualized in Figure 3-19.

Figure 3-19. *Travvy now has all the required information for the booking*

At all times, Harold should be able to close/stop the conversation. Therefore, in the top-right corner of the screen, there is a close icon.

Test the Conversational Interface

Evaluating the dialog in an early stage of the design process gives you the opportunity to improve the flow before a lot of expensive development effort is done.

Figure 3-20 shows that good design saves money.

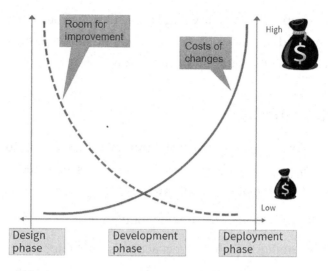

Figure 3-20. *Good design saves money*

On the Internet, you can find all kind of tooling to mock up your design. Figure 3-21 represents an example of such a mock up.

Figure 3-21. *Mockup of a possible start screen for Travvy on an iPhone*

By creating a mockup, in a very early stage of the design process, you can textually fine-tune the answers and questions of your bot and find out what the best way is to present your users the options you want to offer them.

Cognitive Walkthrough

With the mockup, you can do a cognitive walkthrough, a usability evaluation method in which you work through a series of tasks – which the user is expected to carry out – and ask a set of questions from the perspective of the user like

- Will the user try and achieve the right outcome?

- Will the user notice that the correct action is available?

- Will the user associate the correct action with the outcome he expects to achieve?

- If the correct action is performed, will the user see that progress is being made toward the intended outcome?

These insights are then used to improve the usability of the conversational interface. In Figures 3-22, 3-23 and 3-24 you see the mockup for Travvy, showing different ways of displaying the options for Harold to select.

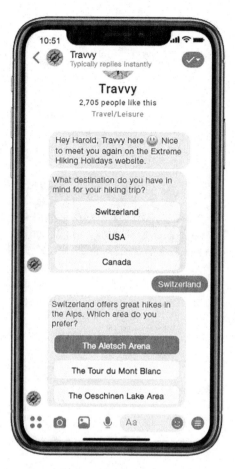

Figure 3-22. *Harold can select destinations just by clicking the preferred option*

Figure 3-23. *In a carrousel, Harold can select destination and hiking level in one go*

Additional to the options for a user to choose, he should be able to formulate/type/ say his own answer too.

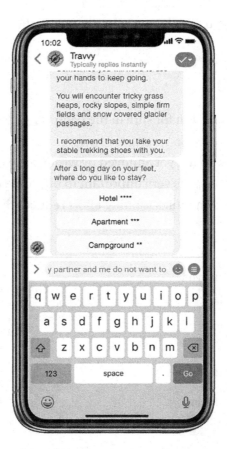

Figure 3-24. *Harold types his own answer*

Usability Test

You can use the mockup to do a usability test with representative users like Harold, just by asking them to perform some predefined tasks, interact with Travvy and then observe his actions.

If you do this with the just developed real conversational interface, you do not only test its usability, but in one go you also train Travvy to better respond to the users' requests, questions, and answers.

To get an overall idea about how the user feels about Travvy, you can ask him the following:

- Mention three things you liked best about Travvy.

- Mention three things you did not like about Travvy.

- How can we improve Travvy?

The answers will help you to enhance your digital assistant.

Train the Chatbot

You can ask Travvy everything, but for sure in the beginning, Travvy will **not** understand everything you're asking.

To make sure that users do not get annoyed by wrong answers, or by Travvy not understanding their questions, it is very important to put some effort in training Travvy.

Figure 3-25 shows that companies training effort reduces the moment the digital assistant gets smarter.

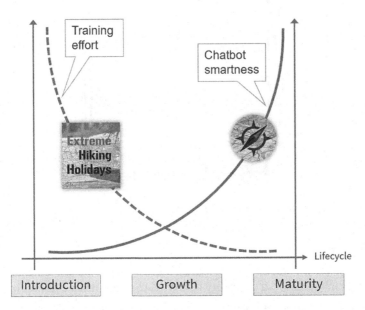

Figure 3-25. *Training effort reduces when Travvy gets smarter*

In the beginning, training will cost time. But training is simple, and you will see very soon that Travvy gets smarter.

Training in Three Steps

Travvy best can be trained with utterances provided by users. However, you do not want to confront real users with an untrained/ignorant chatbot. Therefore, it is better to do the training in three steps:

1. Train the chatbot yourself. Use Travvy by entering as many utterances as possible. You found out a lot about Harold; you know how he might react toward Travvy.

2. Ask the reservation employees of Extreme Hiking Holidays (the stakeholders) to try out Travvy with their knowledge of the customers who normally call/email them.

3. Use the usability test (as explained in the former paragraph) to gather utterances (entered by real users) and check Travvy's reply.

To avoid Travvy asking for clarification of the user's question or, even worse, giving an inappropriate answer, set the **confidence threshold**[1] to 0.35 or higher and the **win margin** to 0.2. (on how to do that, see Chapter 9).

Then the chatbot will first check which intents match with the question with a more than 35% chance. If more intents fulfil that requirement, the bot will look at the difference in probability between those intents. If the difference is less than 20%, then it will display both or more intents as options for the user to choose from.

While training and testing the chatbot, you can find reports of the results in the Insights of the application. In the Retrainer tab, you can see which utterances lead to an unresolved intent (Travvy does not know the answer). You can then decide if you want to add these utterances to a certain intent or maybe even create a new intent for them.

---

[1]Only the top intent that exceeds the confidence threshold is picked if its confidence score exceeds that of other intents by this value or more. If other intents that exceed the confidence threshold have scores that are within that of the top intent by less than the win margin, these intents are also presented to the user (minimum value 0, maximum value 1).

PART III

Implementing Your Digital Assistant

CHAPTER 4

Where Design and Implementation Meet

Introduction

In the previous chapters, you learned all about a chatbot's design and how Extreme Hiking Holidays designed their Digital Assistant, Travvy. The actual implementation of this Digital Assistant can be done by developers by coding all the flow logic, but Oracle Digital Assistant (ODA) offers a feature that can be used by both business users and designers to set up the initial version of a Digital Assistant after which the developer can take over to finish the implementation. With this feature, the **Conversation Designer**, you can quickly build a proof of concept for a conversational design concept without any need to code OBotML, nor do you need to create intents or entities. The Conversation Designer allows you to focus on the user experience by setting up a conversation between the bot and a user. After the conversational user experience is created as explained in Chapter 2, the Conversation Designer can be used to create a prototype/mockup of the Skill Bot. From that mockup, the actual Skill Bot can be generated by the click of a button. Then the chatbot can be trained and tested. Figure 4-1 shows where the Conversation Designer fits in the design and development process.

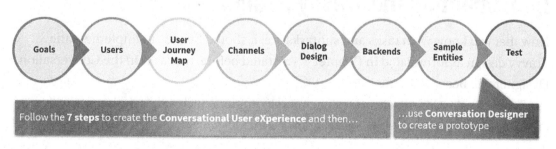

Figure 4-1. *Where the Conversation Designer fits*

© Luc Bors, Ardhendu Samajdwer, Mascha van Oosterhout 2020
L. Bors et al., *Oracle Digital Assistant*, https://doi.org/10.1007/978-1-4842-5422-6_4

In order to use the Conversation Designer, you must have some basic understanding of the concepts used by the Conversation Designer. First there is the concept of tasks. Tasks represent your main use case (Find Trip, Pay Trip, Get Confirmation). The first message in a task should be a message from the user that clearly expresses the selected task/intent. This is needed because when the bot is generated from within the Conversation Designer, the user message will be used to create variants of it as utterances. The intents will be derived from the tasks.

As an example, the first message could be "I want to find a trip." This message is imperative and ends with a noun. This helps the Conversation Designer to classify this message as an Intent Utterance.

Next there are subtasks. These subtasks typically perform supporting functions and can also be referred to as secondary tasks. A user will not explicitly ask the bot to perform such a subtask.

Note   In version 19.1.5 of Oracle Digital Assistant, the Conversation Designer was introduced as beta release (Figure 4-2).

Figure 4-2. *Beta release*

Implementing the Travvy Design

Now that you know what tasks and subtasks are, it should be easy to implement the Travvy design as formulated in Chapter 4. As stated before, the tasks in the Conversation Designer will become intents.

To invoke the Conversation Designer and to implement Travvy's design, we will create a brand-new and empty Skill (Figure 4-3).

Note   For this example, a new empty skill is created. Using the Conversation Designer on an existing skill is not a good idea. This is because existing Intents, Entities, and code will be overwritten when the Skill Bot is generated from within the Conversation Designer. This process will be explained later in this chapter.

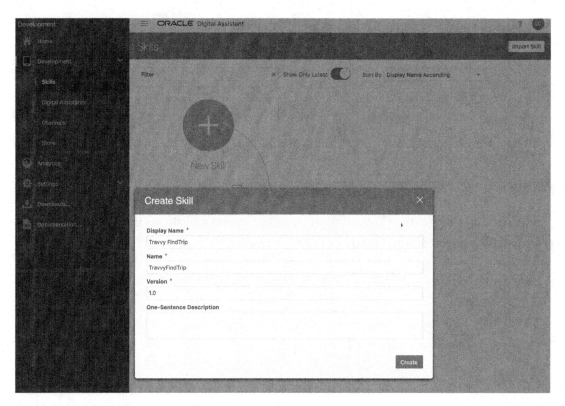

Figure 4-3.   *Create a new skill*

After this new skill is created, you will notice the "Conversation Designer" button in the left-hand toolbar. By invoking this button, you will be guided through the creation of a Skill Bot.

The welcome screen enables you to add tasks that your Skill Bot should be able to do. You can enter as many tasks as you need (Figure 4-4), and at a later stage, you can add more if needed.

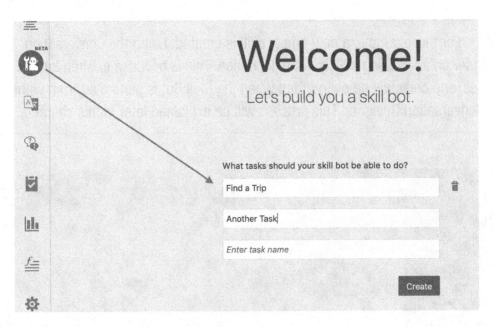

Figure 4-4. *The Conversation Designer welcome screen*

By clicking **Create**, the Conversation Designer opens and shows you the tasks (Figure 4-5) that you entered on the welcome screen.

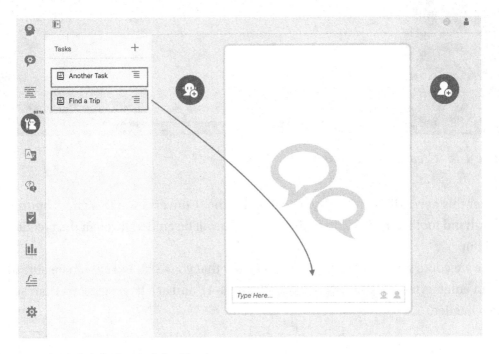

Figure 4-5. *Initial view of the Designer*

With this empty skeleton in place, you can now start creating the mockup of the bot's conversation.

For the purpose of this book, not all the steps involved in creating the mockup for this skill will be explained in this section, but the core functionality will be discussed in detail.

As you can see from Figure 4-6, there are two icons in the Conversation Designer:

When this icon is clicked, a user utterance can be entered.

When this icon is selected, a bot reply can be entered.

When you only need to enter simply text, there is a shortcut that can be used. Simply type the text in the conversation box at the bottom and click the icon (bot or user) to add to conversation.

Figure 4-6. *Adding simple steps*

As explained before, the first message in a task should be a message from the user that clearly expresses the selected task/intent so that reasonably good utterances can be generated from it. In this case, for the **FindTrip** task, for example, *I want to go on an extreme hiking trip* can be entered. The next step is to add a reply of the bot. A simple reply where the bot introduces itself is what we need here. Something like: *Hey Harold, I'm Travvy the digital assistant for EHH. Great to meet you again here.*

Now the conversation start is set up, and we are ready to add some real content to the conversation. As described in Chapter 4, the FindTrip Skill (Figure 4-7) consists of several steps, which we will implement (not all) in this chapter. Let's start with *Indicate Holiday Destination.*

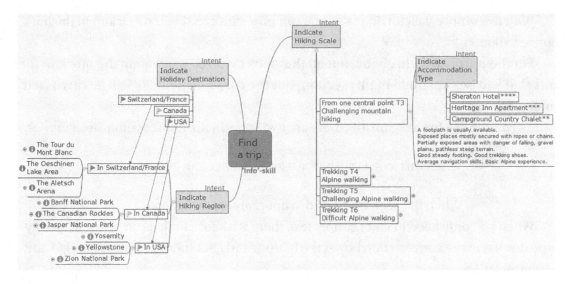

Figure 4-7. *Dialog flow to find a trip*

Indicate Holiday Destination enables the user to pick the favorite destination, and after that the user will be guided to the available hiking trips at that destination. If you read the previous sentence again, you might notice that it contains the following:

- An action to pick the destination

- A branch that routes to the favorite destination

- A subtask to pick one of the available hiking trips at that destination

This is exactly what the Conversation Designer allows you to enter. By clicking the bot icon, an "add message" dialog will be opened that allows you to enter the bot's reply as an "action." In the "actions" section, simply add a message and add as many actions as you like. As indicated in Figure 4-7, the options for the user are

- USA

- Canada

- Switzerland/France

For each of these, an action will be added.

All of these actions will need to branch into the correct "favorite destination" subtask. You could manually create these subtasks, but the Conversation Designer allows you to do this automatically. If you invoke the hamburger menu next to the actions textbox (Figure 4-8), you can select the action type (GoURL or Branch Conversation). In this case, we want to branch the conversation, so that is what should be selected. When you select that option, the following textbox allows you to select an existing (sub-)task or create a new one by entering a nonexisting name.

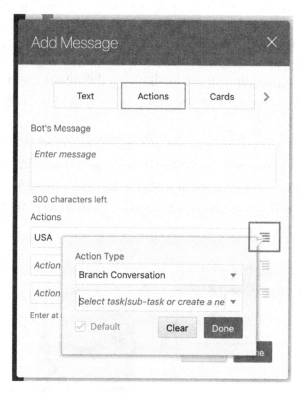

Figure 4-8. *Branching the conversation*

For each of the "Branched Conversation" actions that are created (Figure 4-9), the Conversation Designer has automatically created a new subtask.

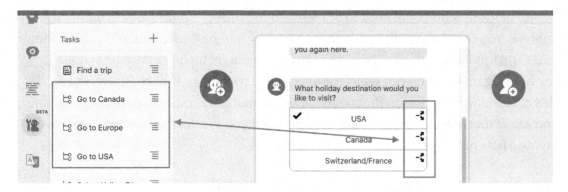

Figure 4-9. *Branches and subtasks*

The next step in implementing the design is to define the content of the subtasks. This works exactly the same as "regular tasks" such as the FindTrip that was previously described.

As an example, the "Go to USA" subtask will be implemented. Where the FindTrip task used Actions to have the user pick the destination, in the "Go to USA," we will use Cards.

Cards are a bit fancier UI component that also allows for images. You can enter image URLs, and below that you can add actions for the user to pick. An example is displayed in Figure 4-10.

Figure 4-10. *Cards and images*

Now you can preview individual tasks and subtasks of the Skill Bot in the Conversation Designer. This preview mode – activated by clicking the "Preview your Task" link in the top-right corner of the screen – allows you to check the flow by playing the currently displayed individual (sub)task. It is not a fully functional representation of the Skill Bot and only plays back the conversation exactly like it was entered in the Designer. This means that clicking actions or scrolling through the options of a carousel with cards is not possible. The main goal is to "preview" your Skill Bot.

Once you are satisfied with the mockup, the final step of the conversation design is to generate your Skill Bot, train it, and test it.

As soon as you invoke the **generate** button at the top right of the screen, a popup opens (Figure 4-11) that warns you that you are about to generate the Skill Bot. When you click "Generate," the magic is invoked, and the Skill Bot is generated.

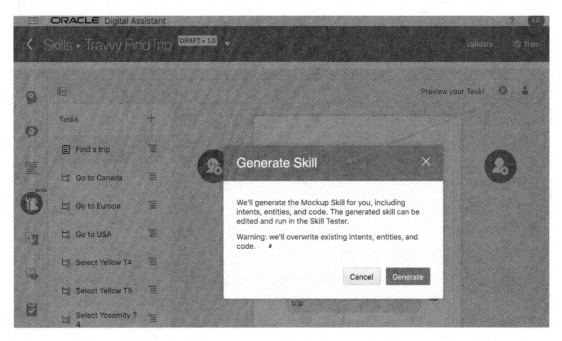

Figure 4-11. *Generate Skill*

Note   Before you can use the Skill Tester, at the bottom of the left-hand menu to test/use the Skill Bot, you must train it by clicking "train" in the blue bar at the top.

Reviewing the Generated Artifacts

To understand how the Conversation Designer works, it is best to have a closer look at the generated artifact. This will allow us to know where Intents and Entities are extracted from the mockup and how the dialog flow is derived. First, we will have a look at the Intents. As stated before, Intents are derived from tasks, and the sample utterances are generated based on the first message in the conversation. For the example of FindTrip, the message "*I want to go on an extreme hiking trip*" was entered. Looking at Figure 4-12, you learn the following:

- FindTrip intent is created based on the task Find a Trip.

  - There are five utterances added to this intent, all directly derived from the message as entered in the Designer.

- Greeting intent is created.

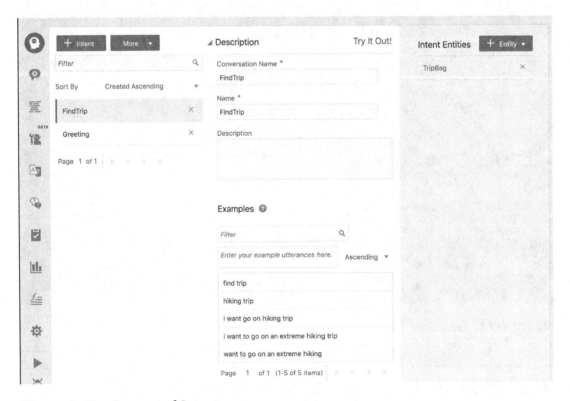

Figure 4-12. *Generated Intents*

Next we can look at the generated entities. As you can see, there are ten entities created for this Skill Bot. These entities (Figure 4-13) are derived from the bot replies as entered in the Conversation Designer. You notice that there are individual entities and composite bag entities, both of which will be explained in more detail in Chapter 5.

Figure 4-13. *Generated Entities*

For example is the "**TripDestination**" entity. This was generated as a "**ValueList**" entity with *USA, Canada, and Switzerland/France* as values. Clearly this was derived from the message *"What holiday destination would you like to visit?"* as displayed in Figure 4-14.

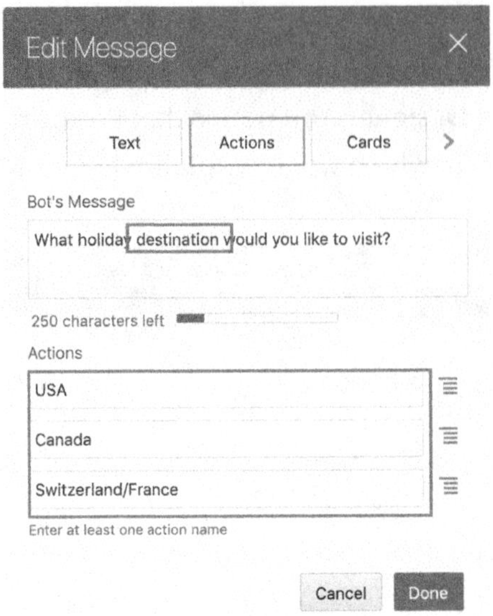

Figure 4-14. *Message used to derive "TripDestination" entity*

Finally, the complete OBotML flow definition YAML file was generated. We will not discuss the contents in detail, but a high-level explanation should be enough to give you a clue. When you open the flow editor, you will notice that the first part contains a whole bunch of lines starting with #! These lines contain the mockup definition as a JSON object.

Note   Sometimes you do not see the lines starting with #! Usually this is because the lines are collapsed, indicated by {←→}. You can expand the section by clicking the small arrow next to line number 1.

Besides the mockup definition, the YAML file also contains the actual flow definition of the Skill Bot. All the states are generated including an "Unresolved Intent" state. There are System.Output components for each simple bot reply and System.CommonResponse components for the complex replies, like the Actions and Cards. An example for "Go to USA" subtask and its individual cards is displayed in the following. First, a variable name "levelCards" is set to contain all info regarding the cards.

```
#------------------ # Task "Go to USA" --------------- #---------------
  setLevelCards:
    component: System.SetVariable
    properties:
      variable: levelCards
      value:
        np:
          title: Yosemite NP
          imageUrl: "<image URL 1>"
          description: "Steep mountains, great views"
          action1Label: Yosemite - Hiking level T4
          action1Branch: Select Yosemite T4
          action2Label: Yosemite - Hiking level T5
          action3Label: Yosemite - Hiking level T6
        yellowstone np:
          title: Yellowstone NP
          imageUrl: "<image URL 2>"
          description: Amazing geysers in full color
          action1Label: Yellowstone - Hiking level T4
          action1Branch: Select Yellow T4
          action2Label: Yellowstone - Hiking level T5
          action2Branch: Select Yellow T5
          action3Label: Yellowstone - Hiking level T6
        zion np:
          title: Zion NP
          imageUrl: "<image URL 3>"
          description: Matching green grey and orange
          action1Label: Zion - Hiking level T4
          action2Label: Zion - Hiking level T5
          action3Label: Zion - Hiking level T6
    transitions: {}
```

This variable will then be used to extract the information that needs to be displayed in the individual cards. All the nitty-gritty details regarding Entities, Intents, and flow will be explained in the next chapters of this book. For now, it is sufficient to know that these are automatically generated for you and that you have some understanding how they are derived from your mockup.

Testing the Generated Skill Bot

The final step is the actual testing of the generated Skill Bot. The one remaining task, if that was not done previously, is to train the Skill Bot. You can easily see if training is required by looking at the icon next to "**Train**" in the toolbar. If it is an exclamation mark, it means that training is required (Figure 4-15). If it is a checked box, that means no training is required.

Figure 4-15. *Training is required*

If training is required, you need to click "**Train**" in the blue toolbar bar at the top and submit the popup that is displayed. The training will start and usually succeed without any failures. You will notice the exclamation mark will change into a checked box, meaning that training was successful. You are now ready to test the Skill Bot. The test can be started by clicking the play button in the toolbar on the left (Figure 4-16).

Figure 4-16. *Start the tester*

The Skill Tester popup is displayed, and you can start a conversation with the FindTrip Skill Bot by entering a message like "*Hi, I want to go on an extreme hiking trip to the USA*" (Figure 4-17). The Skill Bot will reply and offer you the option to select a hiking trip. Note that USA is already preselected as destination, because the bot recognized it from your utterance.

Figure 4-17. Running the tester with the generated Skill Bot

Next you can browse the National Parks in the carousel and select the hiking level you prefer by clicking it. The conversation will continue, just as defined in the Conversation Designer.

You can always reenter the Conversation Designer to make changes to your prototype and regenerate the Skill Bot.

Note   Always keep in mind that regenerating means that all changes that were manually done in the OBotML to Entities, Intents, and flow will be overwritten.

Summary

In this chapter, you have learned how the Conversation Designer of Oracle Digital Assistant can be used to implement a mockup/the initial skeleton of a Skill Bot. The Conversation Designer is very beneficial in the early stages of development as it will not allow you to implement more sophisticated functionality such as Q and A, translations, custom components, and agent integration. These functionalities need to be implanted after generating the OBotML for the Skill Bot. All of this will be explained in the next chapters of this book, where you will also learn more about Entities, Intents, and conversation flow.

CHAPTER 5

Going Through the Implementation Details

Now that you are familiar with the background of Oracle Digital Assistant, what it does, and how you should design a Digital Assistant, it is time to understand some of the basic concepts. During the course of this chapter, you will be first introduced with the Oracle Digital Assistant (ODA) environment. Then you will start your journey with the Digital

Assistant **Travvy**,   based on the use case as described in Chapter 3. With each section, you will be first introduced with the concept and then implement those concepts with respect to the use case.

Introduction to Oracle Digital Assistant (ODA) Environment

Once the identity domain administrator of your Oracle Cloud environment grants you access, you will receive an email on your registered email address for the registration. Post receiving the email, you need to complete your registration for the environment. Once the process gets completed successfully, you will land on the home screen of your ODA as shown in Figure 5-1.

This screen displays a list of instances in your ODA. As you can see in Figure 5-1, we have created one instance named as **dadev** so far considering as a development environment.

© Luc Bors, Ardhendu Samajdwer, Mascha van Oosterhout 2020
L. Bors et al., *Oracle Digital Assistant*, https://doi.org/10.1007/978-1-4842-5422-6_5

Tip   You should use a naming convention for your instances so you can easily understand what you are looking at. The convention used for this book is <2-letter-acronym-product><3-letter-acronym-environment> leading to **dadev** for a digital assistant development environment.

But this list will get updated as you add any new instance for ODA and that name will appear in the list.

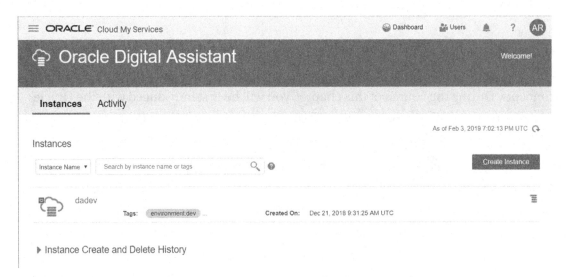

Figure 5-1.   *Oracle Digital Assistant home*

Then to navigate to a specific instance, which in this case is **dadev**, you will have to click the icon ≡ corresponding to the environment and select **Digital Assistant Designer UI**.

Once you select the Digital Assistant Designer UI, you will be redirected to the home screen of that specific instance as shown in Figure 5-2. You will notice that your instance contains a Digital Assistant and few Skills which you didn't create. Those are the samples which Oracle provides you as part of your instance to get you started.

Also, you might have noticed that your home screen contains three tabs:

- All: It is a cumulative view of all the Skills and Digital Assistants your ODA instance contains.

- Digital Assistants: It contains all the Digital Assistant(s) your instance has.

- Skills: All the Skills which your ODA instance has.

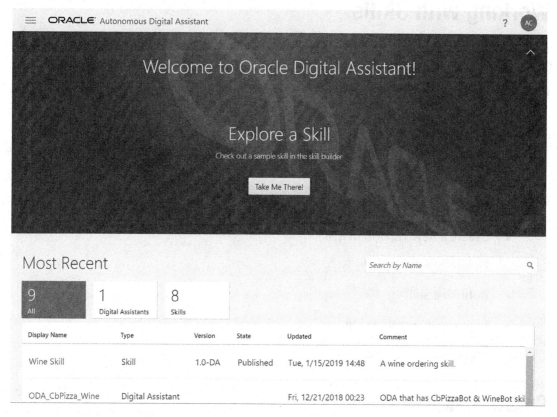

Figure 5-2. *Instance home*

It is important to first understand the difference between a **Digital Assistant** and a **Skill**. You can consider a Digital Assistant as a logical grouping of one or more Skills, Skills which can individually perform specific tasks, to complete an activity. When a user starts a conversation with a Digital Assistant, the Digital Assistant first analyzes the user's inputs, and then, based on the inputs, it takes the user to a specific Skill.

If you consider the use case described in Chapter 4 of this book, Travvy is your Digital assistant. And for Travvy, you will create various Skills (which will be individual chatbots) to perform tasks such as to find a trip, book a trip, update user information, and so on.

Now that you have a basic understanding about Digital Assistant and Skills, next sections of this chapter will take you further into the details.

Working with Skills

As you might have understood already, Skills are the building blocks of any Digital Assistant. Based on your specific scenario, you will create one or more skill for your digital assistant to perform specific task(s). This section will describe how you can

- Create a Skill
- Clone a Skill
- Import a Skill
- Export a Skill
- Create a new version of a Skill
- Delete a Skill
- Publish a Skill
- Export conversation log
- Show channel routes

Create a Skill

Click icon ☰ in the top-left corner of your screen and navigate to Skills under Development as shown in Figure 5-3.

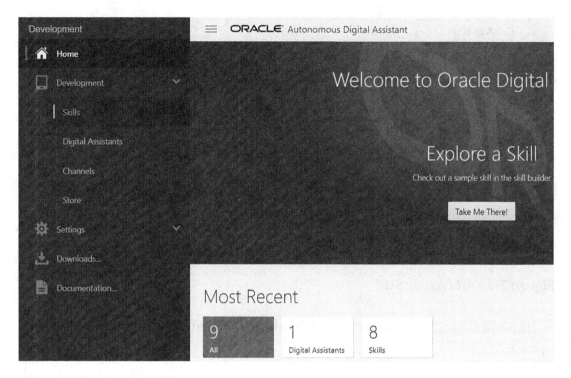

Figure 5-3. *Create Skill*

Once you navigate to Skills, click the icon ≡ to close the side menu.

You will now add a new Skill for Travvy. To do so, you will first have to decide a name, also known as invocation name, for your Skill. There are some guidelines which you should follow when it comes to naming a Skill. Hence, your Skill name should

- Be unique within your Oracle Digital Assistant instance

- Clearly specify the purpose of the Skill

- Be simple, that is, easy to read and pronounce

- Not be a long phrase nor one single word (unless that word explicitly expresses a unique name or purpose)

- Not include words like "open," "go to," or "tell me."

Considering the preceding guidelines, you will name your Skill as **FindTrip**. As the name describes, you will use this Skill to find trips for users of your Digital Assistant.

Click New Skill as shown in Figure 5-4 from the home page of Skills.

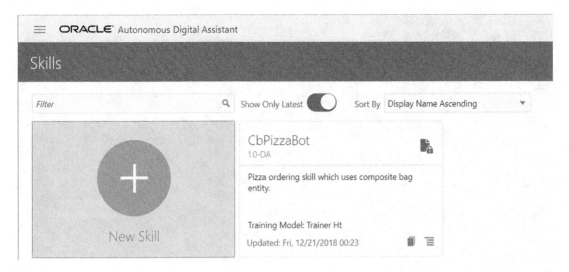

Figure 5-4. *Add a new Skill*

This will open a new window. You need to fill in required details as depicted in Figure 5-5.

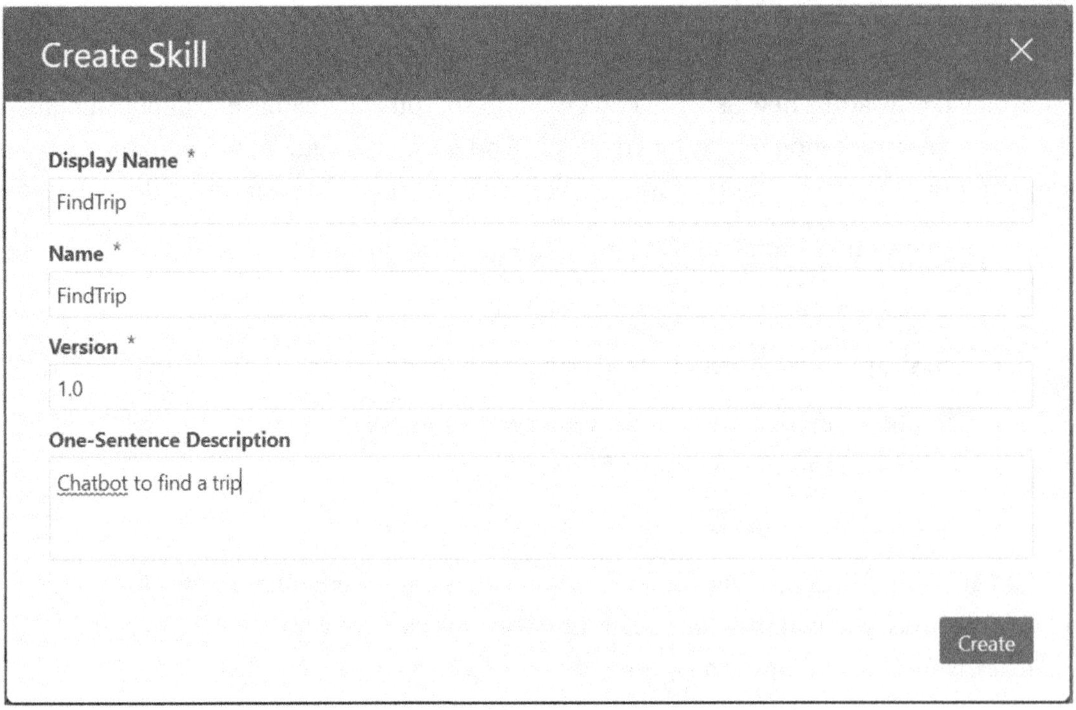

Create Skill   ×

Display Name *

FindTrip

Name *

FindTrip

Version *

1.0

One-Sentence Description

Chatbot to find a trip

Create

Figure 5-5. *Create FindTrip Skill*

Note   As you start entering "Display Name" for your skill, "Name" filed gets auto populated. You can edit this manually, if you wish, to have different "Display Name" and "Name" for your skill.

Once created, FindTrip will appear under Skills section along with other skills. See Figure 5-6.

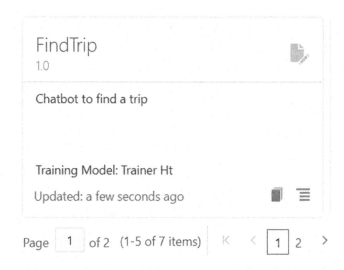

Figure 5-6. *FindTrip Skill*

Click FindTrip skill. This will open a new window of FindTrip skill containing all details which are specific to this skill. Refer to Figure 5-7.

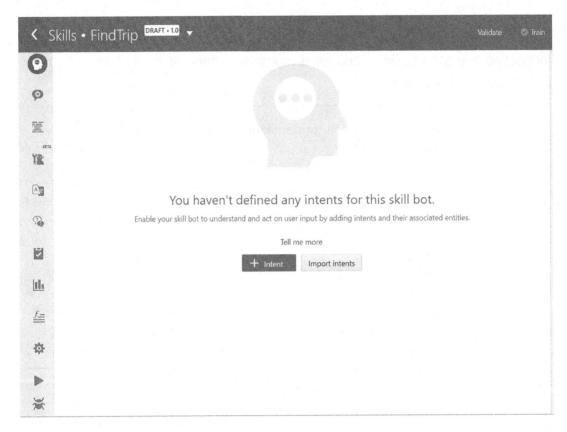

Figure 5-7. *FindTrip skill details*

You can see that top left part of the screen contains navigation menu to take you back to Skills screen.

Moving to the right, you see two menu items, "Validate" and "Train." As the names suggest, they are used to validate your changes and to train your skill.

Left-hand menu items illustrate various sections of your skill. See Table 5-1 which explains these icons with description.

Table 5-1. *Menu Items with Description*

Menu Item	Description
	Intents
	Entities
	Flows
	Conversation Designer
	Resource Bundles
	Q&A
	Quality
	Insights
	Components
	Settings
	Skill Tester
	View problems affecting Skill Bot

Clone a Skill

At some point, you might want to reuse the functionality you designed in a Skill into a new Skill without making any changes to your existing Skill. For this purpose, you can clone your existing Skill into a new Skill.

From the Skills screen (see Figure 5-4), identify the Skill which you want to reuse. Click the Options icon ☰, located at the bottom-right corner of your Skill, to open the Skill catalog as shown in Figure 5-8.

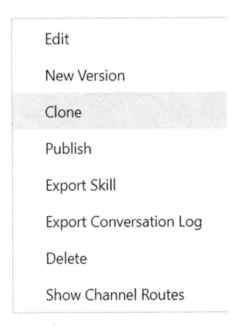

Figure 5-8. *Skill catalog*

From the catalog, select **Clone** option to clone the Skill.

Export a Skill

Another useful feature which you can access from the Skill catalog is **Export Skill**. As the name suggests, by using this feature, you can export your Skill which will be then downloaded on your system. For example, if you export the FindTrip Skill, you will notice a FindTrip(1.0).zip file gets downloaded to your system. Here 1.0 is the version of your Skill.

Import a Skill

If you want to use a prebuilt Skill for your Digital Assistant, you then can import that Skill to your Digital Assistant environment. From the home page of Skills, click the **Import Skill** button as shown in Figure 5-9 to import the zip file containing the Skill.

Figure 5-9. *Import Skill*

Create a New Version of a Skill

From the Skill catalog (see Figure 5-8), select **New Version** if you want to add a new version for your Skill.

Delete a Skill

If needed, you can delete your Skill by selecting the **Delete** option from the Skill catalog (see Figure 5-8).

Publish a Skill

Once you are done with the development of your Skill, you might want to access it from the Digital Assistant. To access a Skill in the Digital Assistant, the Skill first needs to be published. To publish your Skill, from the Skill catalog, select the **Publish** option. Ensure that you train your Skill before publishing. How to do this will be explained in the "Train and Test Intents" section.

After publishing a Skill, you can only make changes to its custom parameter values from the Configuration tab in Settings section.

If you wish to update your Skill post publishing, then you will have to create a **new version** of the Skill.

Export Conversation Log

The purpose of the conversation log is to check various conversations which happen with your Skill. This includes conversations with respect to intents, Q&A (if present in the Skill), and the Skill as a whole. By default, all sorts of conversation logging are turned off

for Skills. You can enable them from the General tab by accessing Settings for a Skill ⚙ .
The following options will be displayed:

Enable Conversation Logging ❓

Intent Conversation

Q&A Conversation

Skill Conversation

Figure 5-10. *Conversation logging options*

Once you enable conversation logging, select **Export Conversation Log** from the Skill catalog (see Figure 5-8). This will show you a new popup window as shown in Figure 5-11. From here, you can select which conversation log you want to export. You can also specify the **Time period** for which you want to get the logs. To finally export the required conversation logs, click the "Export" button.

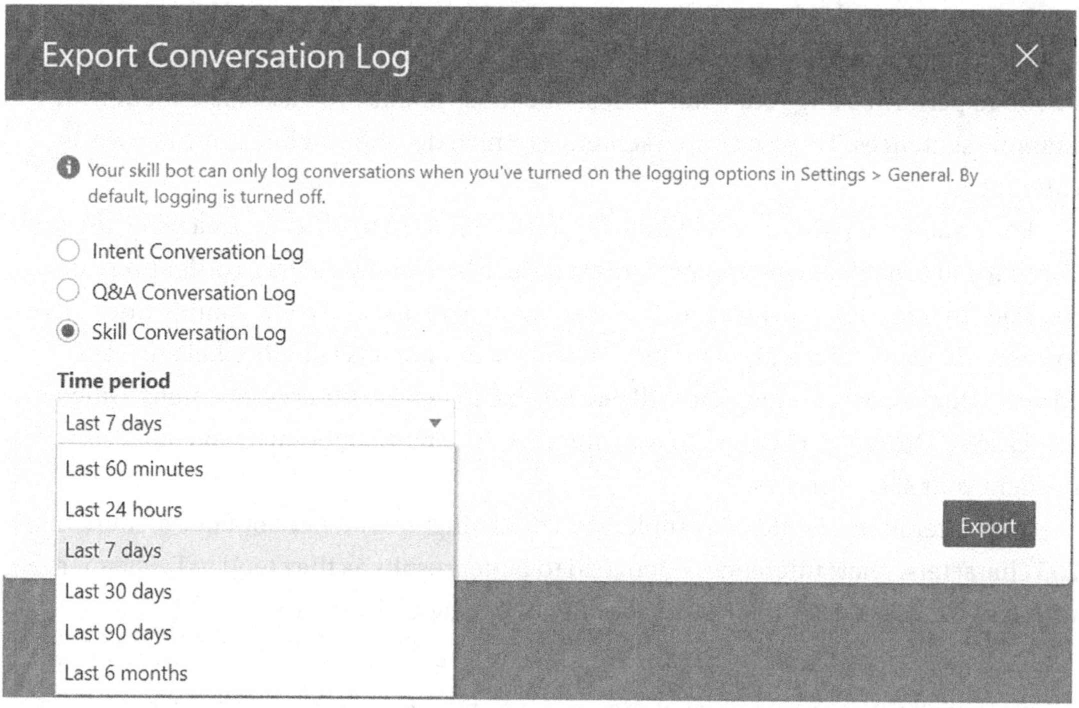

Figure 5-11. *Export conversation log*

Show Channel Routes

The **Show Channel Routes** option from the Skill catalog allows you to route channels to your Skill. You will learn more about channels and how you can route those channels to your Skill in a later section of this book.

Working with Intents and Utterances

In this section, you will learn to

- Create an Intent

- Add Utterances to Intent

- Export Intents

- Import Intents

- Train and test Intents

In Chapter 2 of this book, you were made familiar with Intents and Utterances.

For a quick recap, an **Intent** can be described as a piece of work or task which you intend to perform using your Skill. To describe those intents, you associate them with example sentences. Those example sentences which describe an intent are known as Utterances.

For instance, if you create a PizzaSkill, you might want to order a pizza using the skill. For that you can add an intent, say "OrderPizza," which allows users to order pizza using the skill. To make the skill aware of the user intent, you just add a few sample utterances to your skill like "Order a pizza for me." When a user interacts with the skill and says/ writes "Order a pizza for me," the skill will be able to resolve the specific intent, which in this case is "OrderPizza," based on the utterance. The more utterances you add, the more resilient your skill becomes.

Your utterances should be complete and meaningful sentences but not of more than 255 characters. Short utterances often lead to better results as they tend to be cleaner with less fill words. Also, utterances should not be made of a single word.

Tip   Initially, you should at least create ten to fifteen utterances for each intent you define. This ensures better intent resolution. By the time you move to production environment, you will need more utterances. Identifying utterances for an intent is an iterative process that – if done well – leads to good results.

Each Skill must have at least two intents associated with it. So, if we consider the same example for PizzaSkill, then it should contain two intents, say OrderPizza and CancelPizza.

Create an Intent

In this section, you will add intents to your FindTrip skill which you have already created earlier in this chapter. Drill down the FindTrip skill and navigate to the Intent section from the left-hand menu.

Click the "+ Intent" button to add an intent. As you click the button, you will notice an intent gets created with a default name of "Intent1." See Figure 5-12. You will then give a suitable name to your intent which in this case will be SelectTrip.

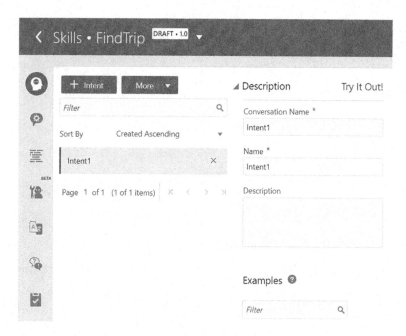

Figure 5-12. *Create Intent*

Update the intent's **Conversation Name** as "Select Your Trip" and **Name** as "SelectTrip." Conversation Name is displayed during conversation by your skill (as an option), when it is unable to resolve an intent, whereas, in your dialog flow, you identify any intent by its Name. If you want, you can also add a **Description** for your intent although it is not mandatory.

Add Utterances to Intent

To add utterances to your intent, you need to add them under the **Example** section, in the text area with place holder "*Enter your example utterances here*" (see Figure 5-13).

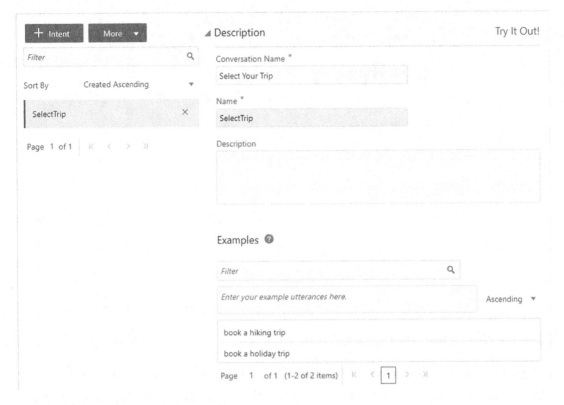

Figure 5-13. *Add Utterances*

Once you start adding utterances, this list will grow. If you want to look for a specific utterance, you can search for it by entering the utterance in the "*Filter*" field which will then issue a search in the preadded utterances.

Export Intents

You can export intents from one skill and then use those intents for another skill. This will save your effort of creating the same intents, if needed, for various skills. The following steps will depict how you can export intents:

1. Navigate to the **Intent** section ⬤ in your Skill.

2. Click the More button and select **Export intents** (see Figure 5-14).

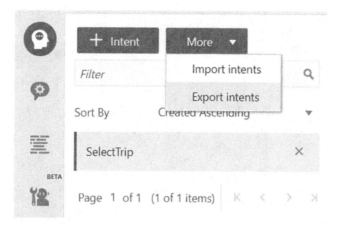

Figure 5-14. *Export Intent*

This will export your intents into a CSV file. In this case, it will be FindTrip-Intents. csv. The exported CSV file contains three columns, namely, **query**, **topIntent,** and **conversationName**. The "query" column contains utterances, whereas the "**topIntent**" and "conversationName" columns contain the name of the related intent and conversation name of that intent (see Figure 5-15).

```
query,topIntent,conversationName
book a holiday trip,SelectTrip,Select Your Trip
book a hiking trip,SelectTrip,Select Your Trip
```

Figure 5-15. *FindTrip-Intents.csv*

Note   When you export intents, all intents present in the skill will be exported into a CSV file.

Import Intents

To import intents into your skill, you need to import the CSV file containing exported intents as shown in previous step. The following steps will depict how you can import intents:

1. Navigate to **Intent** section ![icon] in your Skill.

2. Click the More button and select **Import intents**. See Figure 5-16.

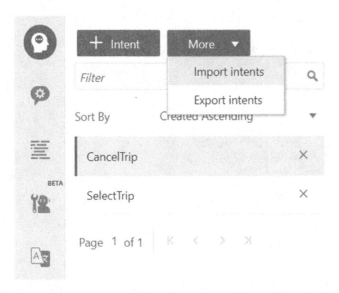

Figure 5-16. *Import intents*

As mentioned earlier, "Export intents" will export all the intents which a skill contains. Hence, "Import intents" will add all the exported intents to your skill. If you want to add any specific intent along with associated utterances, then you need to manually edit the exported CSV file before importing it into your skill.

Train and Test Intents

Once you create or update intents, you need to train your skill so that the model gets trained with your changes. This training is based on different machine learning algorithms which train your skill with the related intents and utterances. When a user interacts with your skill, this training ensures that your skill identifies a specific intent and acts accordingly.

When you add or edit your intents, you will notice an indication mark (⬤) accompanying the Train button in the top-right corner (see Figure 5-17). This implies that you need to train your skill in order to accommodate your changes.

Figure 5-17. *Indication that training is required*

Clicking the Train button reveals two different training mechanisms: **Trainer Ht** and **Trainer Tm** (see Figure 5-18). You can select a specific mechanism by clicking the related radio button. Notice that Trainer Ht is selected by default. During the initial development cycles, when your training corpus is small, use Trainer Ht to train your skill. But as training corpus becomes mature and you start getting accurate results for intent resolution, you should eventually move to Trainer Tm.

For now, you will use the default Trainer Ht training model, to train your skill. As it is the default selection, to start the process, you just need to click the Submit button.

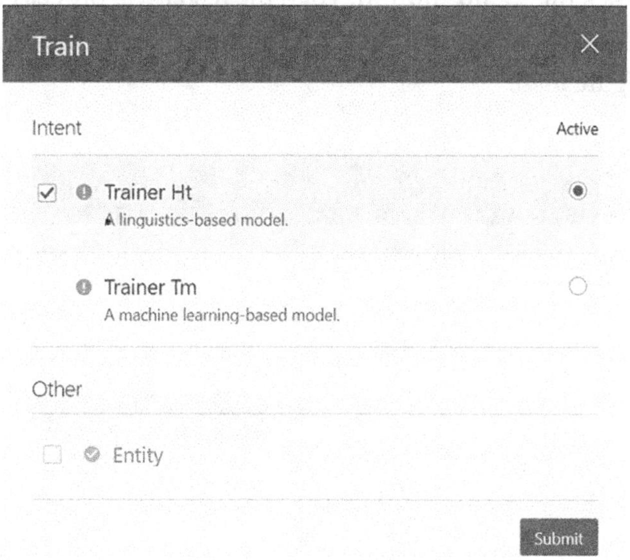

Figure 5-18. *Train Intent*

Presuming that you have trained your skill, next you will test your intent.

Testing is one of the core aspects of any development cycle, and bot testing is in no way different from that. Bot functionalities should be tested as you proceed with its development. Once you finalize the development of intent(s), you should test whether the intents are functioning as they are expected or not.

To test your intents, click **Try It Out!** as shown in Figure 5-19.

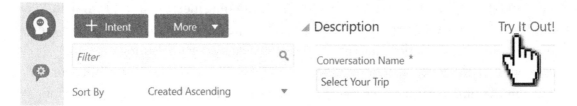

Figure 5-19. *Test Intents*

This will open a new window where you will be able to test your intents. Type a phrase to test and click the "Send" button. Then the tester will process your phrase and show you which intent has been resolved for your phrase, along with other additional information. See Figure 5-20.

Figure 5-20. *Intent testing*

Alternatively, you can also test your intents in a batch mode by toggling the **Batch** button and then by uploading the CSV file. Check Figure 5-21.

Figure 5-21. *Batch upload*

The CSV file you can get when you enable conversation logging for the **Intent Conversation** and then export the conversation log as described earlier in the section **Export Conversation Log**.

Working with Entities

While working with intents and utterances, you learned that intents are specific actions you intend to perform and utterances are the meaningful sentences that describe those intents. Entities are the variable elements which provide context to the intent. In simple words, entities are those values which you seek from a user during a conversation, based on which you will define the outcome or flow of the bot.

For example, in the case of your FindTrip skill, if a user says "I want to book a trip," then you know that the intension of the user is to book a trip, but at the same point in time, it is a very generic statement. Instead of that, if the user says "I want to book a hiking trip to the USA," then the context is much clearer. So, the user wants to book a trip, but to which location? And the answer is to the USA. This location, which helped to provide an appropriate meaning to the intent, is known as the entity. Hence, if the user provides you a location, you can book a specific trip for him, or else you need to ask him the location so that you can give him the appropriate option to book the trip.

Entities are classified in two categories:

1. Built-in entities

2. Custom entities

111

Built-in or system entities are the ones which are provided to you by Oracle as part of Digital Assistant instance. Custom entities are those entities which you will create based on your requirements.

Built-in Entities

Built-in entities are divided into two categories:

1. Simple entities

2. Complex entities

Simple Entities

Simple entities are based on primitive types such as strings and integers. Examples of simple entities are

- NUMBER

- EMAIL

- YES_NO

Complex Entities

Complex entities, on the other hand, are based on specific properties. Table 5-2 shows you complex entities and their properties.

Table 5-2. *Complex Entities*

Entities	Properties
ADDRESS	<ul><li>city</li><li>houseNumber</li><li>road</li><li>originalString</li></ul>
DATE	<ul><li>originalString</li><li>date</li></ul>

(*continued*)

Table 5-2. (*continued*)

Entities	Properties
TIME	• hrs • mins • secs • "hourFormat":"PM"
DURATION	• startDate • endDate • originalString
SET	• minute: The range is {0–59} • hour: The range is {0–23} • dayOfTheMonth: The range is {1–31} • monthOfTheYear: The range is {1–12} • dayOfTheWeek: The range is {0–6}, starting with 0 as Sunday • year
CURRENCY	• amount • currency • totalCurrency
PHONE_NUMBER	• phoneNumber • completeNumber
URL	• protocol • domain • fullPath

Custom Entities

Custom entities are the skill specific entities which you will create to store or retrieve your custom values. For instance, in case of a financial bot, you can think of AccountType entity which can have values such as Savings and Current accounts. Or for a PizzaBot, CrustType can be Thin, Thick, or Medium.

The following are the types of custom entities:

- Derived: As the name suggests, they are derived from built-in entities or custom entity defined by you.

- Value list: These are the entities based on a predefined list of values.

- Entity list: These entities are a super set of entities and based on the list of other entities.

- Regular expression: These entities are based on regular expression (regex) to strictly match to specific pattern.

- Composite bag: These entities are group of related entities which can resolve to specific domain or specific function. You can configure Composite Bag entities in different ways. Either you can give the user prompts to enter values for each and every entity constituting the composite bag, or you can also give the user a prompt to enter a specific entity and derive remaining entities in the composite bag based on provided entity value.

Creating an Entity

To create an entity, you need to drill down the Skill in which you want to add the entity and then navigate to Entity section. In the next few steps, you will add TripLocation entity to your FindTrip skill:

1. Click the FindTrip Skill.

2. Navigate to Entities ( ) from the left-hand side menu.

3. Click the "+ Entity" button.

This will add a new Entity with default name as Entity1. You will then update it for TripLocation. Check Figure 5-22.

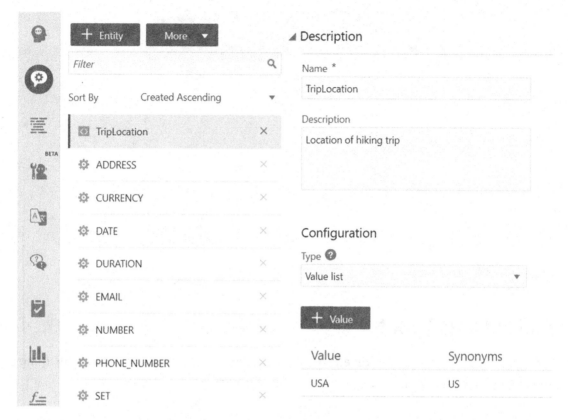

Figure 5-22. *Create Entity*

You can add new values to the list by clicking the "+ value" button. This will then give you a popup as shown in Figure 5-23, from where you can add new values to the list.

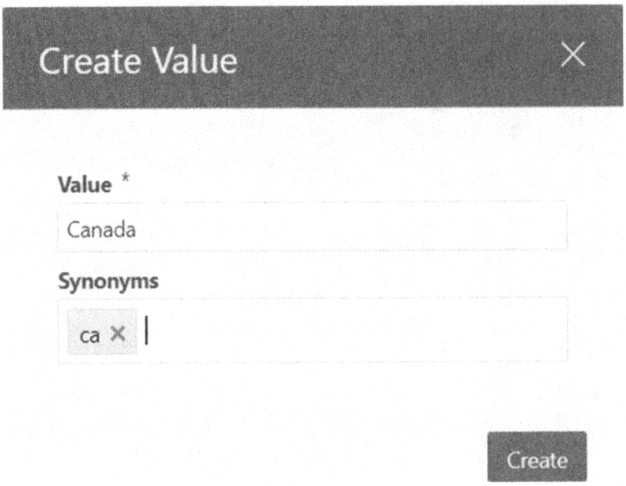

Create Value   ✕

Value *

Canada

Synonyms

ca ✕ |

Create

Figure 5-23. *Add new value*

Add Entity to Intent

Once you create the TripLocation entity, you will then add the newly created entity to the SelectTrip intent. To do this, just follow the following steps:

1. Navigate to **Intent** section ( ).

2. Select "SelectTrip" intent.

3. Click the "+ Entity" button on the right side.

4. This will open a popup containing all available entities.

5. Click "TripLocation" which will then add the entity to "SelectTrip" intent.

Refer to Figure 5-24.

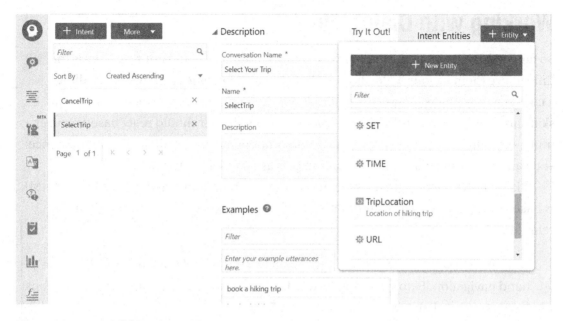

Figure 5-24. *Add Entity to Intent*

As you have just updated the intent, ensure that you train the model and test your intent as described in previous section of this chapter. Additionally, if you have added an entity, the Entity option should be selected while you train your model (Figure 5-25).

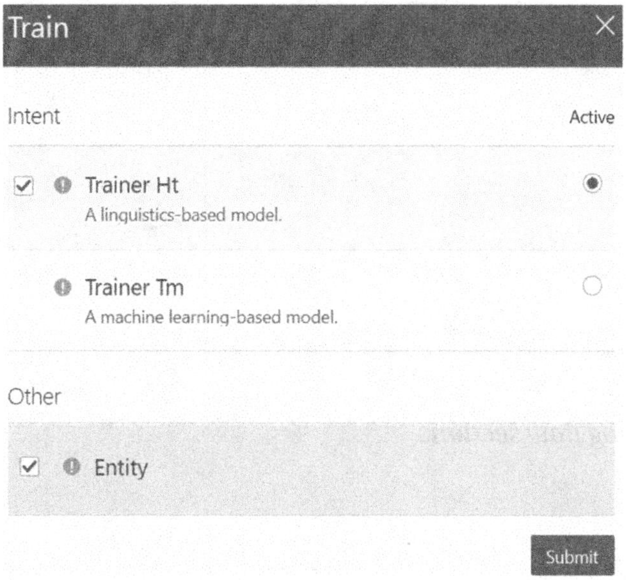

Figure 5-25. *Train Skill*

Working with Dialog Flow

So far you have been made familiar with intents, utterances, and entities. Now is the time to use them in your dialog flow so that your skill becomes capable of handling a conversation. Dialog flow is the place where you craft a skill's conversation. How a skill should greet, what it should ask from users, and how it should react based on user input – everything is handled using various components in your dialog flow. And hence, you can say that the dialog flow definition acts as a model for any skill.

Dialog flows are designed using Oracle Bots Markup Language (OBotML). OBotML is based on YAML, which is a human-friendly way of expressing data and data relationship.

You can access the dialog flow for your skill by simply clicking Flows (⬚), from left-hand navigation. Before moving forward, it is crucial to understand various sections which constitute a dialog flow. Refer to Figure 5-26.

Figure 5-26. *Dialog flow sections*

A dialog flow can be broadly classified into three sections:

1. Header: Which contains OBotML platform version and dialog flow name.

2. Context: Contains context and variables, also known as dialog flow variables, which are being used in the flow.

3. States: Contains individual states, actions, and their outcome. This section contains the main logic based on which a skill interacts and responds.

As OBotML follows YAML standards, it is important to follow the indentations throughout the dialog flow. In case you write a line without proper indention, you will receive errors during the autosave which happens during the design time.

Note   Each indent is equal to two blanks in length.

Header section of a flow is pretty state forward. It contains the metadata information of your bot. But when it comes to context and states, you should have a look at it more closely.

Context Section

As mentioned before, you define the global scoped variables for your skill in context section which you can then use across various states in your skill's dialog flow. A variable in context section is defined as

variableName: "**variableType**";

As obvious, variableName is the name which you give to your variable. variableType, which is the type of variable, can be

- Primitive types (string, int, boolean, double, and float)

- Built-in entity (CURRENCY, ADDRESS, etc.)

- Custom Entity (like TripLocation created in previous sections)

- nlpresult, which is the result from the natural language processing engine

- resourcebundle, which is used for the variable you define to reference resource bundles in a skill

These variables always get flushed when a conversation ends or when they are explicitly getting resetted.

States Section

States is the places where you define the entire flow for a skill. Here, you define all activities and actions that your skill is supposed to perform. Your flow, in general, comprised of many states. These states invoke component, which do the task or work for the state.

Note   A state can invoke one and only one component.

A flow always goes from top to bottom. What this means is that in states section, the first state which you define first gets executed first, the second state gets executed next, and so on. This is the default behavior unless you explicitly set the transition to a different state.

Also, it is highly recommended to have a default transition in your flow to gracefully handle any unexpected exception/error which can arise due to an unforeseen user input.

There are two more things which you should know before you start working with dialog flows. They are Apache FreeMarker Template Language (FTL) and components in the dialog flow.

First, let's tackle Apache FreeMarker Template Language and how it works.

Apache FreeMarker Template Language

To keep things simple, in context of ODA, you use Apache FreeMarker Template Language (FTL) to set and access values of variable (and their properties, if any). If you are interested to know more about FTL, you can check their official web site https:// freemarker.apache.org/.

Syntax for using FTL in dialog flow is pretty straight forward. You wrap everything for which you want to use FTL inside ${...}. You use FTL inside the property definition of components (components are described in next section of this chapter). For example, if you have defined a variable as myVariable and you want to read its value, then you simply use

- ${myVariable.value}

Or if you define a variable, which is a complex entity of type ADDRESS, say userAddress, then you can use FTL as follows to extract the city:

- ${userAddress.value.city}

Use of FTL is not only limited to this. You can also use FTL to write complex expressions in your flow. A very commonly used scenario is using FTL for System. CommonResponse components where you define conditions with "if," "else," or "elseif" directives. The syntax for this is

- <#if yourCondition>result1<#else>result2</#if>

- <#if firstCondition>result1<#elseif secondCondition>result2<#esleif thirdCondition>result3</#if>

Of course, the use of FTL is not limited to CommonResponse components, and you can use them throughout your dialog flow for assignment of values in your bot response or evaluation purposes.

In addition, you can also use various operations which FTL supports for strings, numbers, date-time values, and so on. To know more, refer to the link `https://freemarker.apache.org/docs/ref.html`.

A prolong discussion on FTL, covering all such use cases and operations, may go well beyond the scope of this book. Hence, we are concluding this section here. Any further uses of FTL will be explained as needed.

Components in the Dialog Flow

Any functionality which you want to implement in your skill comes from components. Whether you want to display cards or a list, perform a specific action based on a user input, display any output, state transition, handle error, and so on or even if you want to perform some specific task like invocation of a REST call to process user response, you will always do all these with the help of components.

Components can be broadly classified in two types:

- Built-in components

- Custom Components

As the name suggests, built-in components are those components which Oracle provides you as part of the Digital Assistant framework, to use in your skill. These components are sufficient to perform most of the skill's actions. You can easily identify these components in your flow as they are always preceded by "System." in "states" section under "component." Refer to Figure 5-26.

At any point, if you want to perform some action in your skill which cannot be done using built-in components, then you can define your own components. Those components, which you create to perform specific tasks, are known as custom components.

Built-in Components

As mentioned, these are predefined components in ODA framework. To add a built-in component into your flow, you need to click the "+ Components" button on top of the editor. This will open a new window as shown in Figure 5-27 providing you various component types to choose.

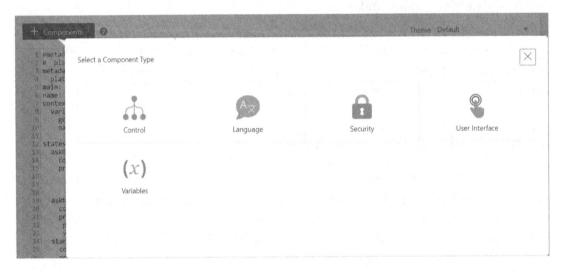

Figure 5-27. *Component types*

As you can see in the preceding figure, ODA provides you five types of built-in components. By selecting a component type, you can select a component template which falls under that type. In the following, you can find a complete list of component templates which fall under each component type shown in Figure 5-27:

1. Control

   i. component: "System.ConditionEquals"

   ii. component: "System.ConditionExists"

   iii. component: "System.Switch"

2. Language

   i. component: "System.DetectLanguage"

   ii. component: "System.Intent"

   iii. component: "System.MatchEntity"

   iv. component: "System.QnA"

   v. component: "System.TranslateInput"

   vi. component: "System.TranslateOutput"

3. Security

   i. component: "System.OAuth2AccountLink"

   ii. component: "System.OAuthAccountLink"

4. User Interface

   i. component: "System.AgentConversation"

   ii. component: "System.AgentInitiation"

   iii. component: "System.CommonResponse"

   iv. component: "System.List"

   v. component: "System.Output"

   vi. component: "System.ResolveEntities"

   vii. component: "System.Text"

   viii. component: "System.Webview"

5. Variables

   i. component: "System.CopyVariables"

   ii. component: "System.ResetVariables"

   iii. component: "System.SetVariable"

To add any of these types of components along with its whole state, you need to click the type which will then give you the option to add a specific component of that type. Refer to Figure 5-28 which you get when you select Control type. All you need to do is to select the "Apply" button to add the component in your flow.

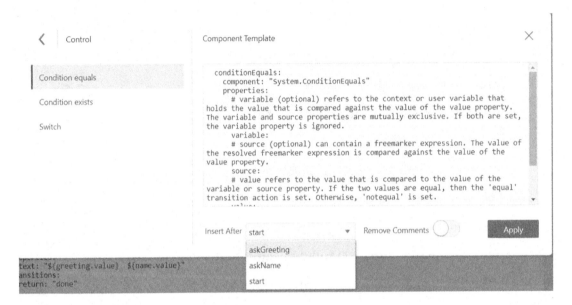

Figure 5-28. *Adding Control type component*

In the preceding figure, you can see that when you add a built-in component, the framework provides you with a prebuilt code snippet which you then update based on your requirement. If you notice, "Insert After" dropdown allows you to select after which state in the flow you want to add your component. Similarly, the "Remove Comments" toggle lets you add the component with and without comment.

You will see the use of these components as you progress with this book. Some components such as the ones for Q&A and user agent will be covered in individual chapters to make you familiar with them. Other generic components will be explained based on their use, whenever feasible.

Custom Components and Webview

When you have specific requirement for which built-in components are not sufficient, then you create your custom components. Custom components are implemented as service in your skill. After you define the custom component service for your skill, you can refer them in your dialog flow.

Consider a case of accessing data using an API based on user input, where you want to retrieve some user-specific detail. Such a need cannot be fulfilled with built-in components, and you will write your own custom logic to implement such scenario in a Custom Component. You refer to a custom component in your flow as

- component: "YourCustomComponentName"

You write your custom component code in JavaScript. For your ease, Oracle provides you SDK to implement your custom components. Custom components can be implemented in three different ways:

1. Deploying a custom component in the embedded container in the Oracle Digital Assistant

2. As an API using Oracle Mobile Hub (Oracle Mobile Cloud)

3. As an external Node service

To add a custom component to your skill, navigate to Components ($f_{\equiv}$)section from left-hand navigation menu. Select Custom tab (which is selected by default). Refer to Figure 5-29.

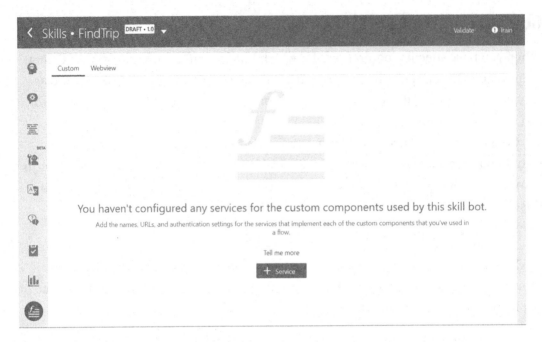

Figure 5-29. *Adding Custom Component*

Then you need to click the "+ Service" button which will give you abovementioned options to implement your custom component. Refer to Figure 5-30.

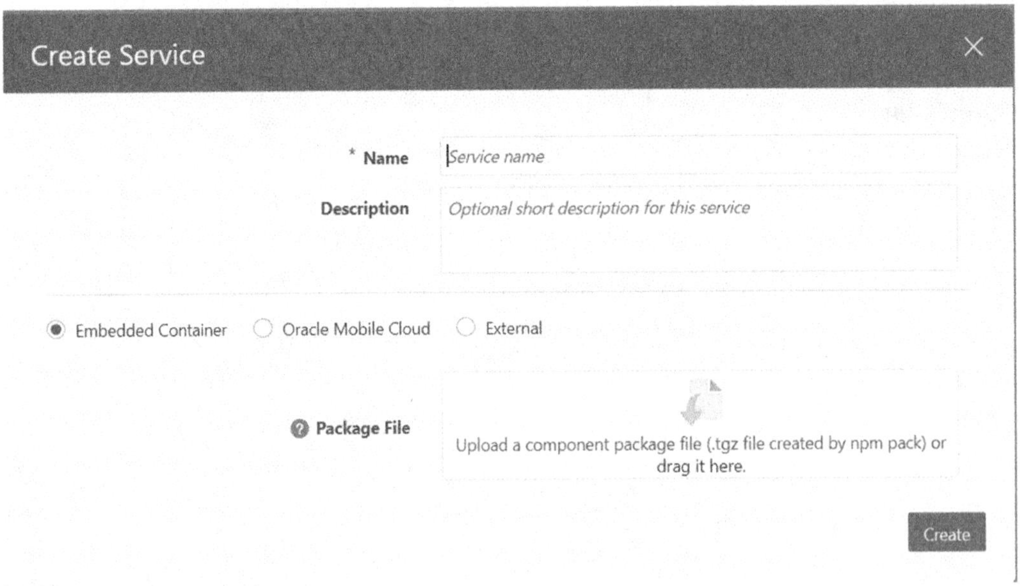

Figure 5-30. *Create Custom Component service*

Custom components will be explained in detail in Chapter 10.

Refer to Figure 5-29. Beside Custom tab, you have another tab as Webview. Click the Webview tab. Here you can create Webview components for your skill. As the name suggests, Webview allows you to integrate web pages to your skill. By clicking the "+ Service" button on Webview tab, you can create Webview service for your skill as shown in Figure 5-31.

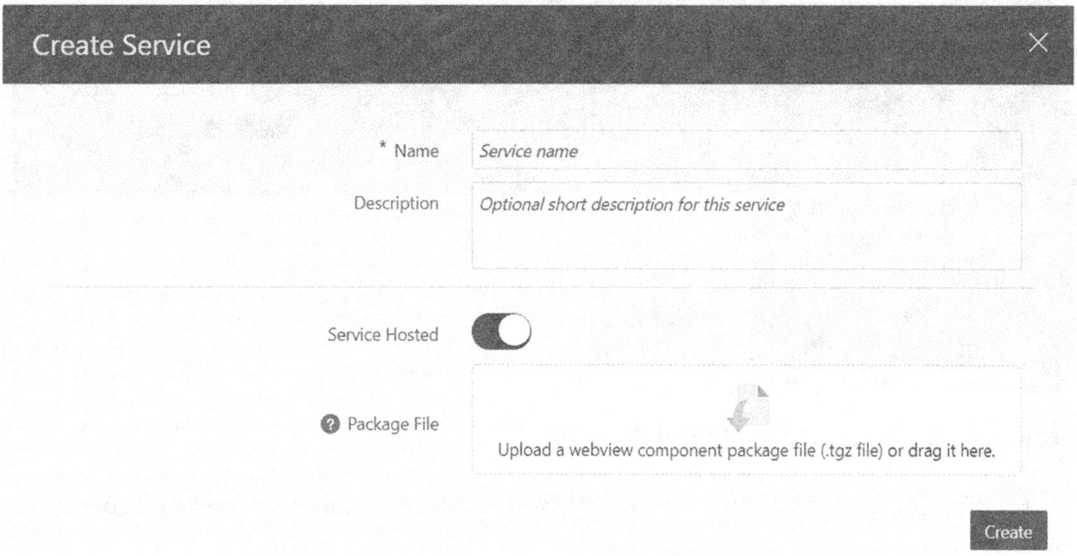

Figure 5-31. *Create Webview service*

Webview components are discussed in detail in Chapter 8.

Adding Skill to a Digital Assistant

As explained earlier in this chapter, you add various features to a Digital Assistant using Skills. Once skills with specific features are ready, then you publish those individual skills and add the published skills to Digital Assistant. Also, a skill must be published in order to add it to Digital Assistant. Considering that your skill **FindTrip** is ready and published, this section will focus on how you can add **FindTrip** skill to the Digital Assistant **Travvy**.

Navigate to the Digital Assistants section of the welcome screen of your Oracle Digital Assistant by clicking the Menu (≡) icon as shown in Figure 5-32.

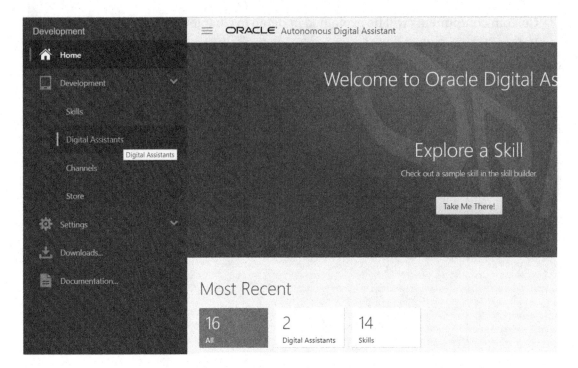

Figure 5-32. *Navigate to Digital Assistants*

From your Digital Assistant screen, click New Digital Assistant as shown in Figure 5-33.

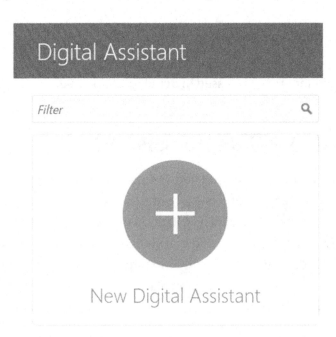

Figure 5-33. *Add new Digital Assistant*

This will open a new popup for you. Add **Display Name** as "Travvy" and **Description** as "The Digital Assistant for Extreme Hiking Holidays" in the popup and click the **Create** button. Once created, you will be redirected to the Digital Assistant Details screen (refer to Figure 5-34), and this will also add a new section of Travvy on the Digital Assistant home screen.

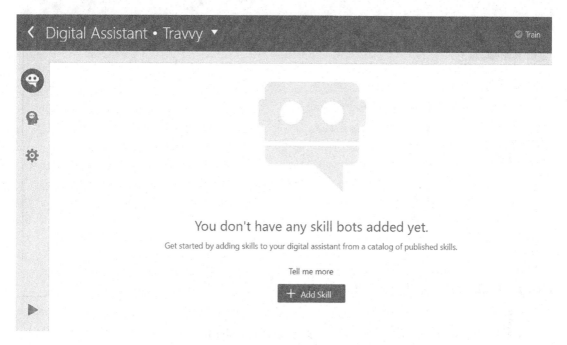

Figure 5-34. *Digital Assistant Details screen*

Click the "+ Add Skill" button as shown in Figure 5-33, to add FindTrip skill. This will redirect you to **Skill Catalog** screen. Click Add Skill (⊕) for FindTrip skill. Once done, you will get a confirmation message that skill has been added successfully.

Navigate back to the Travvy. As you have just made a change by adding a new skill to the digital assistant, you will notice an indication (◑) for training. You will have access to the same training model as for the Skill, that is, Trainer Ht and Trainer Tm, Trainer Ht being default. Train your digital assistant with default Trainer Ht model.

Add a few sample utterances to your skill which distinguish one skill from other skills. In this case as well, you will add a few such utterances for FindTrip skill. Finally, the Skill section for your Digital Assistant Travvy will look similar to Figure 5-35.

Note   These utterances are not meant for training your Digital Assistant. Rather, they are displayed in the Digital Assistant help. The utterances that the digital assistant is trained with come from the configured skills.

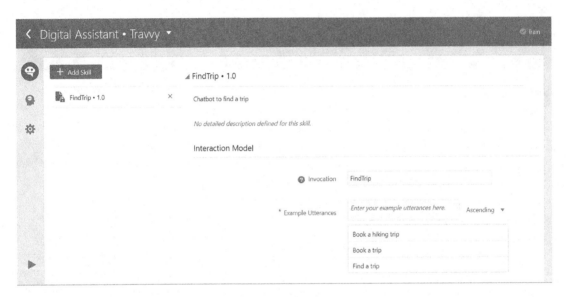

Figure 5-35. *Add Skill to Digital Assistant*

That is how you can integrate a skill to a digital assistant.

Summary

Now that you have come so far, it's the time for a quick recap.

You started this chapter by getting familiar with the Oracle Digital Assistant (ODA) environment, Digital Assistant, and then Skills. Then you took a deep dive into Skills and the underlying concepts of Utterances, Intents, and Entities and how you use them in your Skill. Next step was the dialog flow, where you came to know how you can handle human interaction with your skill and process user inputs with various components available in the Digital Assistant framework to craft your skill communication. And finally, you learned how you can create a Digital Assistant and consume your Skill into it.

We recommend you to have a clear understanding of the concepts discussed during the course of this chapter. In next chapters, we will presume that you are familiar with these concepts and will take you to more advanced concepts of ODA.

PART IV

Enhancing Your Digital Assistant

CHAPTER 6

Exposing Your Digital Assistant over Channels

At the time when you decide to introduce a Digital Assistant in your business, presumably you are aware of your target audience. Being aware of your target audience will assist you in identifying the best possible way to make your Digital Assistant available to them. When you start designing your Digital Assistant or Skill, it is only accessible by you and people who have access to that specific cloud instance. To make your Digital Assistant or Skill available for your business users, you need a channel to expose it to them. Channel, as the name suggests, acts as a medium using which you expose your Skill or Digital Assistant for your end users.

Selecting the right channel to expose your Digital Assistant plays a vital role in its success factor. For instance, if your Digital Assistant or Skill is not easily accessible to end users, they might tend to avoid it and go for a comparatively easier route to accomplish their task. At the same time, if your chatbot is designed to provide user sensitive information, you might want to restrict it to a more secured and closed channel.

Consider a scenario where you would allow your users to perform their day-to-day banking tasks using your Digital Assistant. Tasks such as checking their account balance or make money transfer. In this case, you might want to expose your Digital Assistant over bank's web portal and/or over their mobile applications. This could be because of two reasons. Firstly, a web portal of any bank is considerably more secure. And secondly, because such information and tasks are specific to users of that bank, you don't want to have unnecessary users.

On the other hand, consider a case where you are designing a Digital Assistant for food delivery. To make your Digital Assistant easily accessible as well as popular, you might want to expose it over various other messaging channels in addition to a web channel – channels such Facebook Messenger, Skype, Microsoft Teams, and so on.

© Luc Bors, Ardhendu Samajdwer, Mascha van Oosterhout 2020
L. Bors et al., *Oracle Digital Assistant*, https://doi.org/10.1007/978-1-4842-5422-6_6

In crux, you will always decide which channel you want to choose based on the purpose of your Digital Assistant and your target audience.

Now that you are aware of significance of channel, let's have a look at various out-of-the-box channels which come with Oracle Digital Assistant platform.

Channel Types in Oracle Digital Assistant

Channels in Oracle Digital Assistant are broadly classified into four types. They are

1. Users

2. Agent Integrations

3. Applications

4. System

Users Channel

The Users channel allows you to expose your Skill or Digital Assistant over a user-facing platform. The following is the list of Users channels:

- Facebook Messenger

- Webhook

- Web

- iOS

- Android

- Twilio SMS (text only)

- WeChat

- Slack

- Microsoft Teams

- Cortana

- Skype For Business

Agent Integrations

As it is self-explanatory by its name, you create an Agent Integration channel when you wish to involve a human agent during the course of a conversation. Agent Integration uses Oracle Service Cloud (18C or later) to pass the information provided by a user during the course of a conversation, to a live agent. In that way, when a live agent takes the control of conversation from the bot and starts interacting with the user, they don't need to ask user to provide required information once again. Agent Integration will be discussed in detail in a separate chapter of this book.

Applications

Applications channel is used when you want to have an event-driven conversation. In simple words, you initiate the conversation with your Skill or Digital Assistant at specific (bot flow) state based on an event generated from a different application.

System

This is a system-specific default channel for your Oracle Digital Assistant instance. This channel is used when testing your Skill or Digital Assistant using Skill Tester ▶. You can't create any additional System channel by yourself. It is important to mention here that any changes being made on this channel are applicable across all Skills as well as Digital Assistants in the instance.

Working with Channels

In this section, you will learn to create channel for your Oracle Digital Assistant (ODA). Log in to your ODA instance and click the icon ≡ at top-left corner to open the side menu. Select Channels located under the Development section. Refer to Figure 6-1. This will take you to the Channels screen.

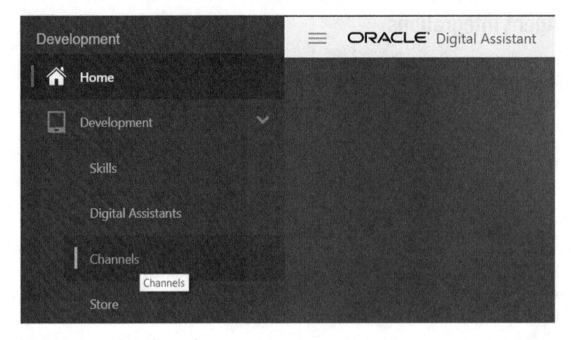

Figure 6-1. *Select Channels*

Once you are on the Channels screen, you will notice it contains four different tabs specific to each channel type discussed in the previous section. By default, the Users tab is displayed when you navigate to Channels as shown in Figure 6-2. It also displays you a list of preconfigured Users channels, if you have created any.

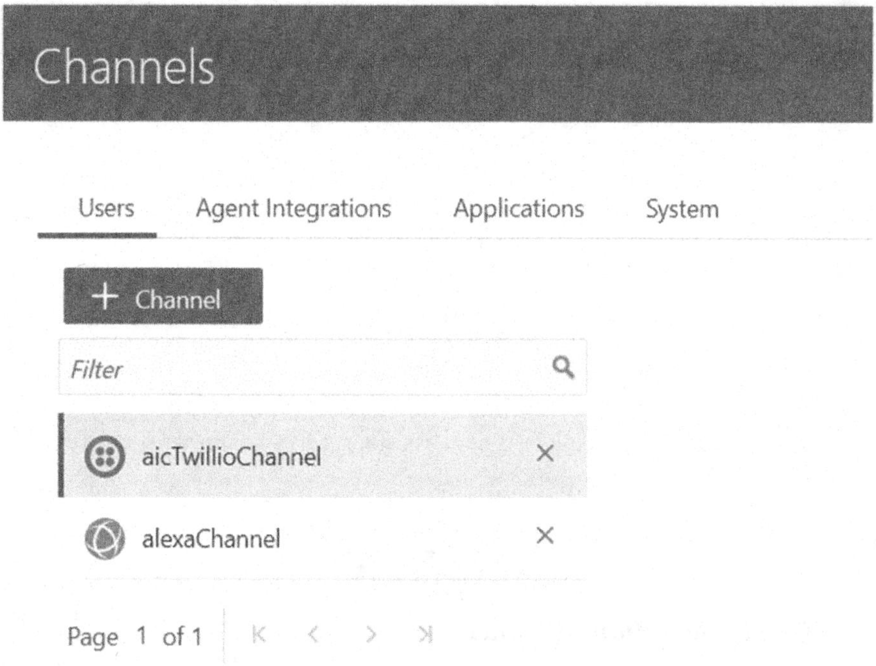

Figure 6-2. *Users channel*

This section allows you to create new Users channels as well as maintain, that is, update, or delete previously created Users channels. This will be discussed in detail in upcoming sections of this chapter.

Next, click the Agent Integrations tab. This tab allows you to create and manage Agent Integrations channels. Similar to the Users tab, this will show you preconfigured channels in case you happen to have any. As you can see in Figure 6-3, we have not yet created any Agent Integrations. Hence, there is no channel being displayed here.

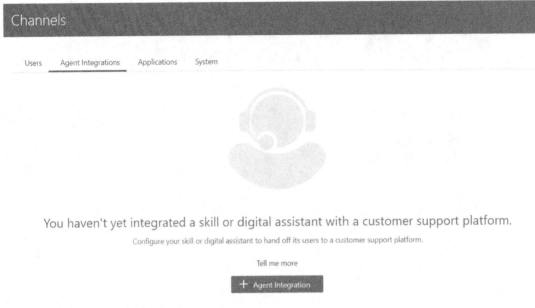

Figure 6-3. Agent Integrations channel

Click the next tab, Applications. This tab allows you to create and manage your Application channel for ODA instance. Similar to previous tabs, you can find all your previously created Applications channel, if you have any. Check Figure 6-4.

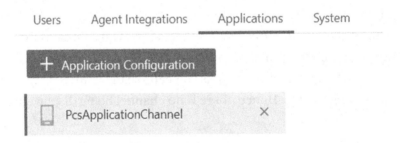

Figure 6-4. Applications channel

Click the last tab of this screen, which is System. As you can see in Figure 6-5, this tab comes with the prebuilt default **System_Global_Test** channel.

Users	Agent Integrations	Applications	System

Channel Enabled		Reset Sessions
System_Global_Test		
Name	System_Global_Test	
Description	Used by the BOT test feature. If it is disabled, the test UI will not work for ANY Bot.	
Channel Type	Test	
Secret Key	7D7D465D62DE0B5CE0537206020A9056	Reset
Session Expiration (minutes)	60 Default	

Figure 6-5. *System channel*

If you toggle Channel Enabled switch on this screen, that would enable or disable all the Skill Testers in your ODA instance based on what you select. Same goes with Reset Session button as well. On click of this button, all Skill Tester sessions will be reset.

Note   Channel Enable switch exists for all channels which you have in your Digital Assistant instance. This switch allows you to enable or disable any specific channel.

Presuming that you are well familiar with various channel types and their significance, in the next section, you will start with creating the Users channel with a focus on the Web channel. Based on our use case of Travvy, we will limit the scope of our discussion to the Web channel in this chapter. Webhook and Agent Integrations will be discussed in detail in a separate chapter, as mentioned earlier. The concept of channel will remain the same for other channel types, but their configurations will differ. We would recommend you to refer to Oracle documentations in case you require to configure other channel types.

Now that the intent is clear for this chapter, let's get back to creation and configuration of the Web channel for Travvy.

Create Users Channel

Click the icon ≡ at the top-left corner and select Development ➤ Channels. From the Users tab, click the [+ Channel] button. This will open the Create Channel window for you as shown in Figure 6-6.

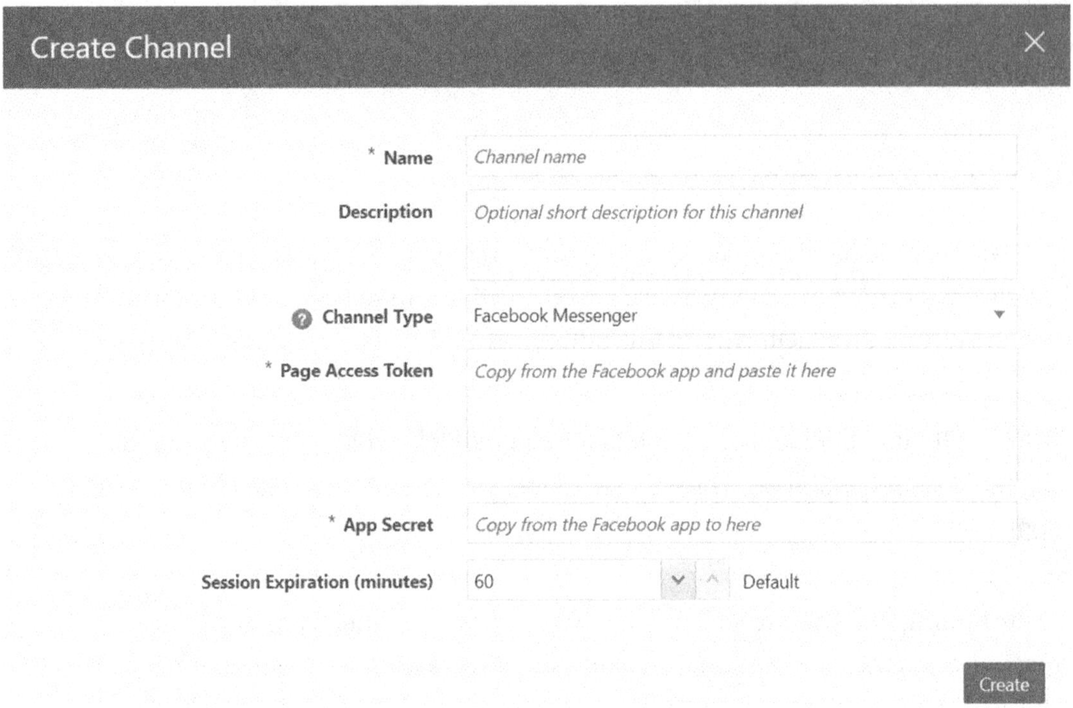

Figure 6-6. *Create Channel*

From here, you will create a Web channel. Change the channel type by clicking the dropdown and selecting Web from the list. Refer to Figure 6-7.

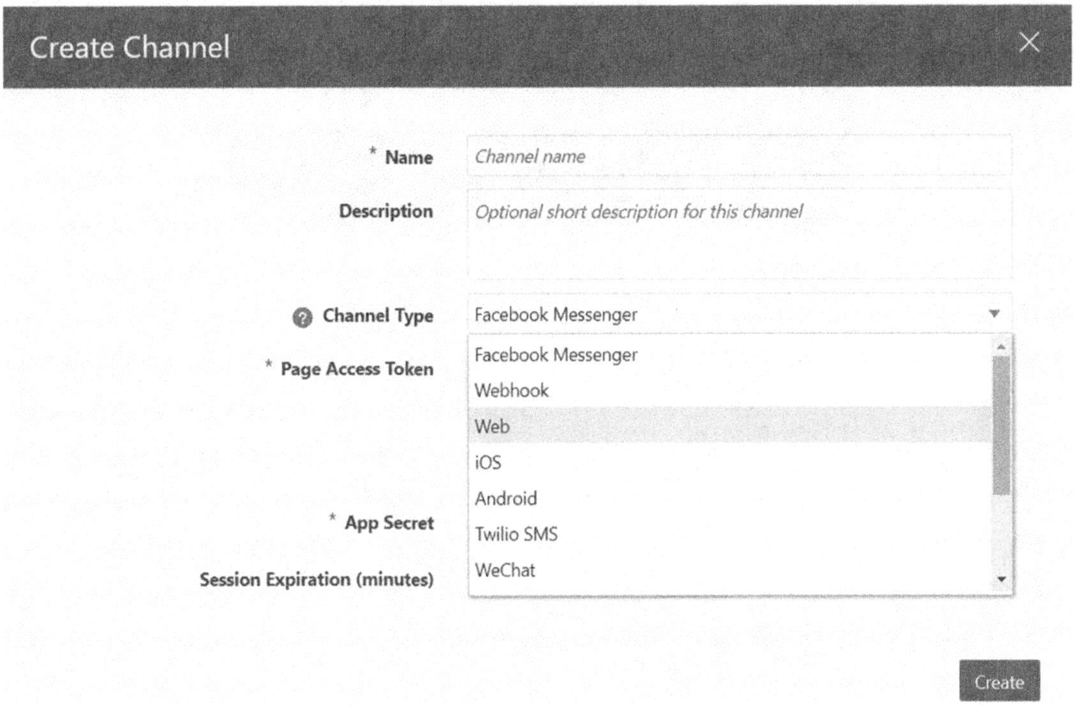

Figure 6-7. *Select Web channel type*

Fill in the necessary details as displayed in Figure 6-8 and click the **Create** button.

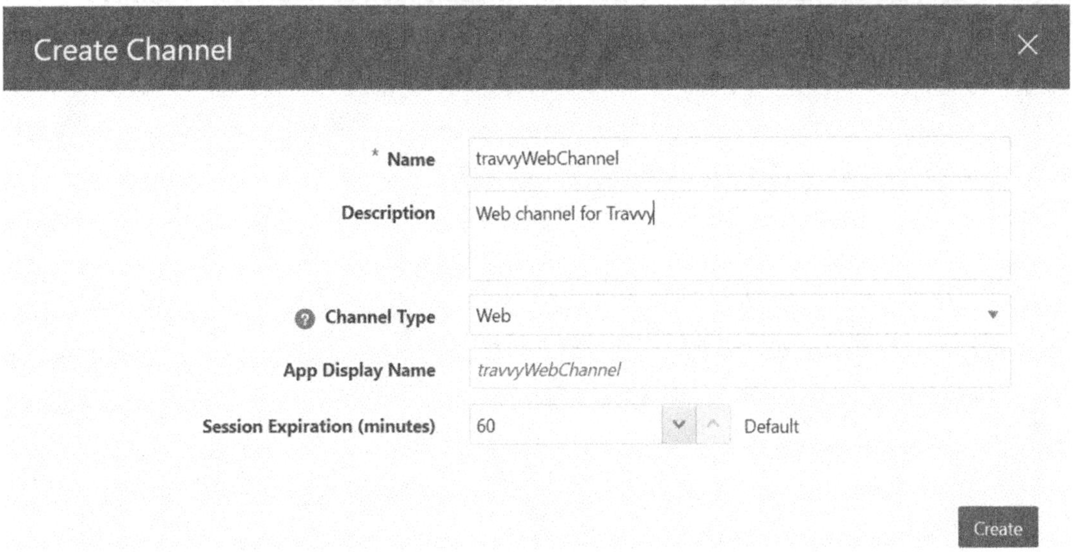

Figure 6-8. *Create Web channel*

Once created successfully, you should be able to see your newly created travvyWebChannel under the list of your Users channel as shown in Figure 6-9.

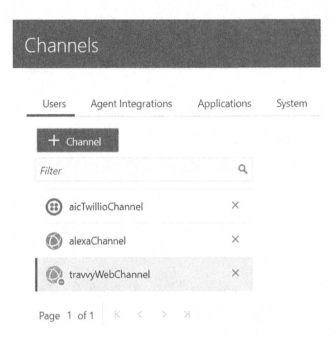

Figure 6-9. *List of Users channels*

By selecting a channel, you will be able to see its details on this screen. Refer to Figure 6-10. This contains various crucial information such as Route To, Channel Enabled, and App Id. You will be using these details in upcoming sections of this chapter.

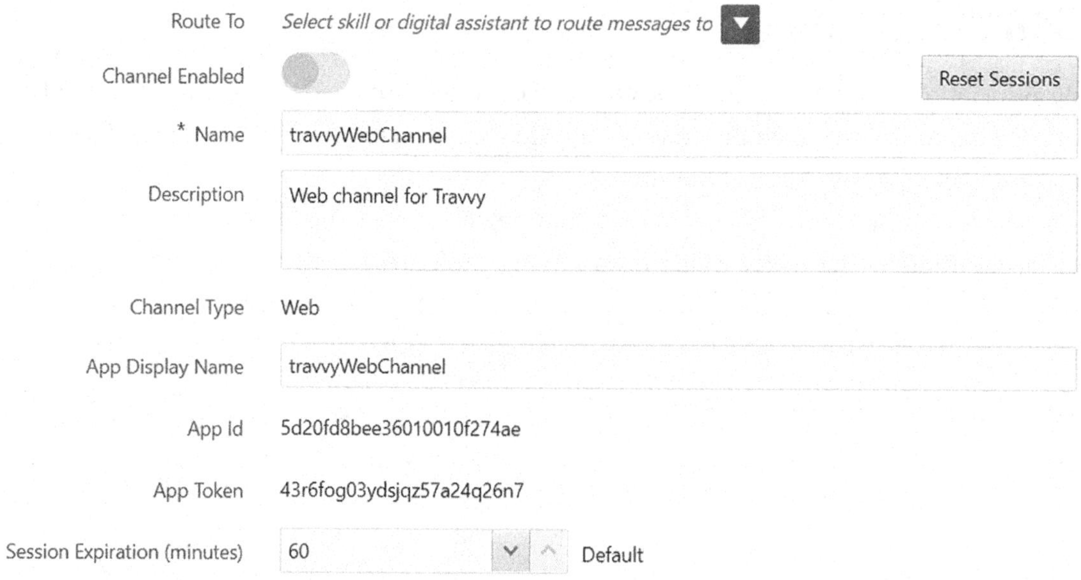

Route To   *Select skill or digital assistant to route messages to* ▼

Channel Enabled   ⬤   Reset Sessions

* Name   travvyWebChannel

Description   Web channel for Travvy

Channel Type   Web

App Display Name   travvyWebChannel

App Id   5d20fd8bee36010010f274ae

App Token   43r6fog03ydsjqz57a24q26n7

Session Expiration (minutes)   60   ⌄ ⌃   Default

Figure 6-10. *Channel details*

As you can see in Figure 6-10, this channel is not yet enabled. To enable a channel, you need to associate the channel with a Skill or a Digital Assistant. The process of associating a Skill or a Digital Assistant to a channel is known as Channel Routing. Channel Routing will be discussed in the next section.

If you could remember from the previous chapter of this book, you always have to publish a Skill in order to associate the Skill to a Digital Assistant. Whereas in case of a channel, you can also associate an unpublished Skill to a channel. This feature comes very handy when you are developing and testing a Skill over a channel, at the same time. And hence, saves you from creating numerous versions of a Skill. So, you can associate any Skill which is published or unpublished or any Digital Assistant to a channel when you need to do so.

In addition, you can also associate a prepublished version of a skill to a channel. This helps you to handle a fallback situation in your production environment, in case needed.

Channel Routing

By associating a Skill or a Digital assistant to a channel, you are defining a specific Skill or a Digital Assistant which the channel should route you to, when it is invoked. This is commonly known as Channel Routing.

Select your newly created travvyWebChannel. Next, click the ▼ button corresponding to "Route To." This shows you a complete list of all available Skills and Digital Assistants you can associate to this channel. Refer to Figure 6-11.

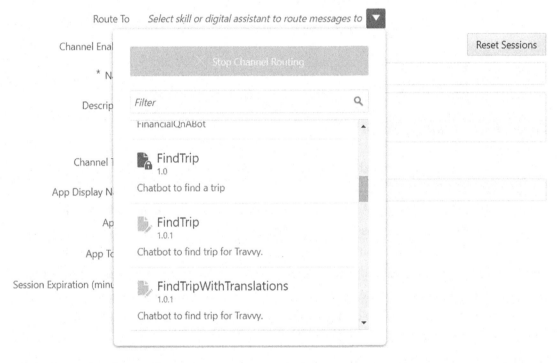

Figure 6-11. *Channel Routing*

As you can see in the preceding figure, a published Skill can be identified by the 🔒 icon, whereas an unpublished will have the ✏️ icon associated to it. Similarly, a Digital Assistant can be recognized by 💬 icon.

In addition, the list also mentions skill's "One-Sentence Description" and Digital Assistant's "Description" underneath the name, which you mention at the time you create them.

Select a Skill or a Digital Assistant from the list which you want to expose. Next toggle the button corresponding to "Channel Enabled" in order to enable the channel. The screen with all details should look like Figure 6-12.

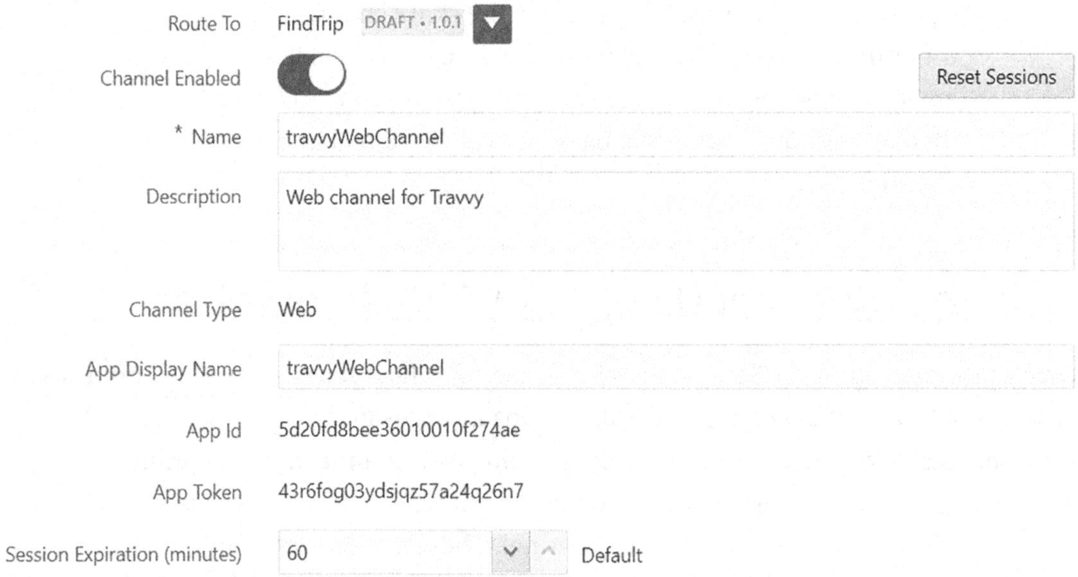

Figure 6-12. Channel details

In order to reroute your channel to a different skill or Digital Assistant, you can simply click the ▼ button corresponding to Route To and change the routing by selecting a different skill or Digital Assistant from the list.

Your newly created Web channel is now ready to be accessed.

Before we proceed further with this chapter, the following are the two points which you should always remember:

1. A channel holds a one-to-one relationship with a skill or a digital assistant. This emphasizes the fact that only one skill or a digital assistant can be exposed at a time over a specific channel. Though, you can create multiple channels for a specific messenger type.

2. When you invoke a digital assistant over a channel, skills associated to the digital assistant will always use digital assistant's channel to send and receive messages. Irrespective of whether skills have their own separate channels or not. In simple words, if you are accessing a skill directly via the skill's channel, then the conversation will take place on the skill's channel. If you are accessing one or more skills via a digital assistant's channel, then all communication will happen over the digital assistant's channel.

Testing a Channel Using ODA Client Samples

Oracle provides you ODA Client samples for JavaScript, which you can download from www. oracle.com/technetwork/topics/cloud/downloads/amce-downloads-4478270.html.

Using these samples, you can quickly test your Web channel in bare minimum steps. These samples come with preconfigured Bots Client SDK, which you will later use in your web application to access the Web channel. These sample apps provide you a playground to test and debug your channel. Follow the following steps:

1. Download bots-client-sdk-js-samples-<Version_Number>.zip file using the preceding link.

2. Extract the zip file to a location of your preference.

3. Once extracted, you will receive two sample applications, namely, chat-sample-web and chat-sample-custom-web. To start working with these samples, you need to have Node.js installed on your system. If not, download it from https://nodejs.org/en/ download/.

4. Navigate inside chat-sample-web directory and run command

   npm install

   using a command prompt/terminal to install application dependencies.

5. Once dependencies get installed, run the following command to start the application:

```
node server.js
```

6. By default the application will be listening to port 3000 of your machine. Open a browser and navigate to URL `http://localhost:3000/`.

You should be able to see a web page as shown in Figure 6-13.

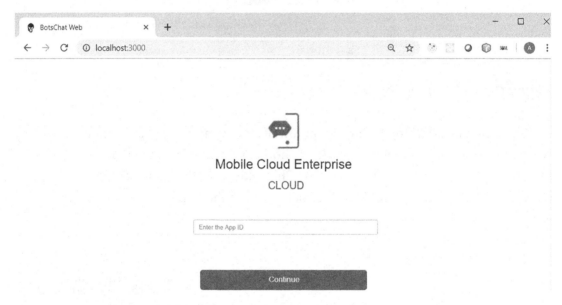

Figure 6-13.   *chat-sample-web*

You need to enter the App ID of your Web channel. You can find this on your channel details section as outlined in Figure 6-14.

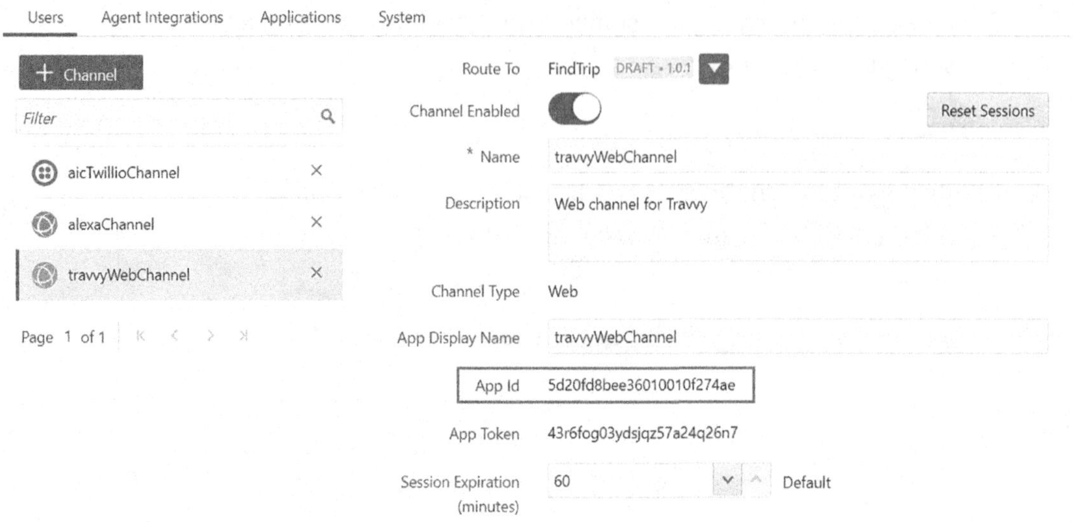

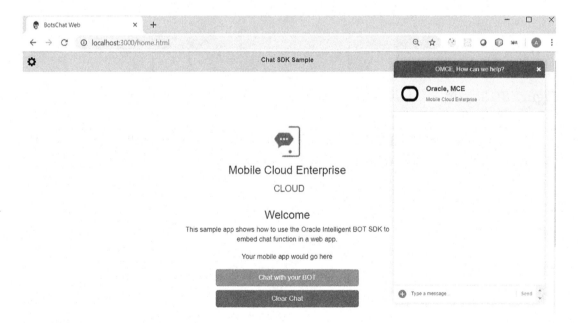

Figure 6-14. *Channel App ID*

Enter the App ID in your browser window of chat-sample-web application and press "Continue" button. This will take you to the home page of the application.

From the home page, click the button "Chat with your BOT." This will open a chat widget, which will invoke your Web channel based on the App ID. Refer to Figure 6-15.

Figure 6-15. *Invocation of chat widget*

You should be now able to chat over your Web channel. Check Figure 6-16.

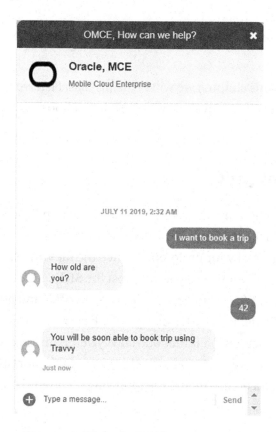

Figure 6-16. *Conversation over Web channel*

Now that you have successfully tested your application, you will use ODA Client SDK for JavaScript in the next section to invoke the channel from your web application.

Accessing Users Web Channel from Web UI Using ODA Client SDK

The following are the steps which you need to perform in order to access your Web channel from a web application:

1. Configure client SDK and host it on a web server.

2. Create a simple web application for Extreme Hiking.

3. Add JavaScript code in the HTML file of web application to invoke client SDK.

4. Deploy the web application and test conversation over the Web channel.

For the purpose of this chapter, we will be using GlassFish web server to host client SDK and the web application. However, you are free to choose any web server of your choice.

Configure Client SDK

Download the latest version of ODA Client SDK for JavaScript from the URL `www.oracle.com/technetwork/topics/cloud/downloads/amce-downloads-4478270.html` and extract it to a location of your preference. Once the files are extracted, you need to configure it for the web server where you will host the SDK.

To start things off, let's first bring up the GlassFish server and add a location where you will put the SDK. By default, a GlassFish server listens to port 8080.

In this case, we have created a folder odaTravvyDemo inside docroot folder on GlassFish server. Inside this folder, you will add a static folder for SDK. In order to configure the SDK, you need to execute the following command on Mac system, as also shown in Figure 6-17.

```
./configure http://localhost:8080/odaTravvyDemo/static/
```

```
Lucs-MacBook-Pro:bots-client-sdk-js-19.1.5.0 lbors$ ./configure http://localhost:8080/odaTravvyDemo/static/
[Done! Files are available in /Users/lbors/Downloads/bots-client-sdk-js-19.1.5.0/http:__localhost:8080_odaTravvyDemo_static_
[Lucs-MacBook-Pro:bots-client-sdk-js-19.1.5.0 lbors$ ▊
```

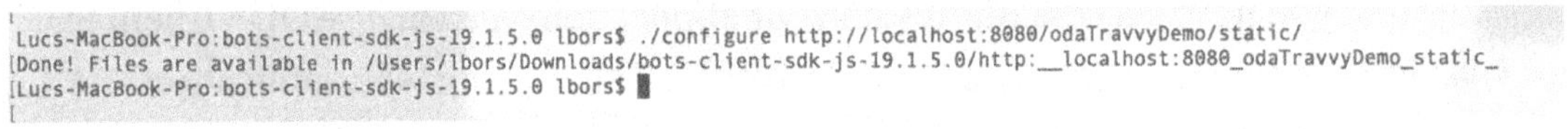

Figure 6-17. *Configure SDK on Mac system*

"localhost" and the port "8080" may change if you wish to configure the SDK for a different web server, based on IP address or in case of any public Internet deployment.

Once you execute the preceding command, it would generate a folder name containing the path you provide to configure. In this case, the folder name will be **"http:__localhost:8080_odaTravvyDemo_static_"**. Rename the folder as static and copy it inside the odaTravvyDemo folder created earlier on the GlassFish server. That's it.

Unfortunately, life is not that simple for people working on Windows machines in this case. Thankfully, you have an alternative. Navigate to js-sdk folder where you have extracted your SDK. Locate loader.json file inside js-sdk folder and open that. You will notice that the file contains

```
{"url":"https://placeholder.public.path/bots.1.16.1.min.js"}
```

Replace "`https://placeholder.public.path`" with "**http://localhost:8080/ odaTravvyDemo/static**" in loader.json. Again, "localhost" and the port "8080" may change if you wish to configure the SDK for a different web server, based on IP address or in case of any public Internet deployment.

Using a text editor, find and replace the same part for each occurrence in remaining files of this folder. For your convenience, the following are the list of files which you need to update in addition to loader.json:

1. bots.1.16.1.min.js

2. frame.1.16.1.css

3. frame.1.16.1.min.js

Note   Version numbers of above mentioned files may change based on SDK version.

Once done, rename the folder as static and copy it inside the odaTravvyDemo folder created earlier on the GlassFish server.

Create a Simple Web Application for Travvy

Now you need to create a web application from which you will invoke the SDK. We have created an Extreme Hiking web application using Oracle JET, but you are open to choose any technology of your choice. Figure 6-18 shows you the Extreme Hiking web site.

Figure 6-18. *Extreme Hiking web site*

Add JavaScript Code

Open the web page where you want to add the widget in a code editor. In our case, index. html is our home page, and we will use Microsoft Visual Studio Code editor.

Add the following script snippet inside the head section of your HTML. This code is responsible to invoke the SDK which you have deployed in previous steps.

```
<script>
    !function (e, t, n, r) {
        function s() { try { var e; if ((e = "string" == typeof this.
        response ? JSON.parse(this.response) : this.response).
        url) { var n = t.getElementsByTagName("script")[0], r =
        t.createElement("script"); r.async = !0, r.src = e.url,
        n.parentNode.insertBefore(r, n) } } catch (e) { } } var o, p, a,
        i = [], c = []; e[n] = { init: function () { o = arguments; var
        e = { then: function (t) { return c.push({ type: "t", next: t
        }), e }, catch: function (t) { return c.push({ type: "c", next:
        t }), e } }; return e }, on: function () { i.push(arguments) },
        render: function () { p = arguments }, destroy: function () { a
        = arguments } }, e.__onWebMessengerHostReady__ = function (t) {
```

```
    if (delete e.__onWebMessengerHostReady__, e[n] = t, o) for (var r
    = t.init.apply(t, o), s = 0; s < c.length; s++) { var u = c[s];
    r = "t" === u.type ? r.then(u.next) : r.catch(u.next) } p &&
    t.render.apply(t, p), a && t.destroy.apply(t, a); for (s = 0; s <
    i.length; s++)t.on.apply(t, i[s]) }; var u = new XMLHttpRequest;
    u.addEventListener("load", s), u.open("GET", r + "/loader.json",
    !0), u.responseType = "json", u.send()
  }(window, document, "Bots", "<your-static-folder-url>");
</script>
```

Once done, just replace the highlighted section of the code with the URL of your static folder. This URL must be the same as the one which you have used to configure your SDK. In our case, it is "**http://localhost:8080/odaTravvyDemo/static/.**"

Once replaced, code should look as displayed in Figure 6-19.

```
<script>
  !function (e, t, n, r) {
    function s() { try { var e; if ((e = "string" == typeof this.response ? JSON.parse(this.response) : this.response).url
  }(window, document, "Bots", "http://localhost:8080/odaTravvyDemo/static/");
</script>

</head>
```

Figure 6-19. *Head section code snippet*

Now you need to add the code snippet to invoke widget from this page. Inside the body section of your page, add the following code snippet:

```
<script>
  Bots.init({appId: '<your-channel-app-id>'})
</script>
```

Replace the highlighted section in the code snippet with the App ID of your channel. Refer to previous section to figure your App ID.

Deploy Web Application and Test

Accommodate preceding changes and deploy your web application. In our case, we deployed the web application in "web" folder. This folder resides at the same level as static folder. Refer to Figure 6-20 to check the deployment hierarchy.

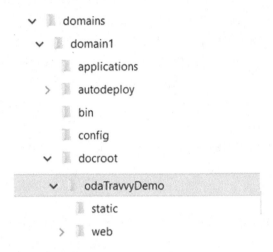

Figure 6-20. *Deployment hierarchy*

After that, access the web application from any browser. If you have followed the exact steps as mentioned before, you should be able to access the application over the URL: `http://localhost:8080/odaTravvyDemo/web/`. Figure 6-21 shows you how the application looks finally. At the bottom-right corner, you can see the chat widget.

Figure 6-21. *Extreme Hiking web application with chat widget*

Click the widget to open it and start testing. Figure 6-22 shows you the conversation using the chat widget from the web application.

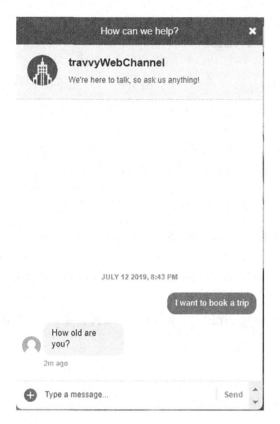

Figure 6-22. *Conversation using chat widget*

Customizing Chat Widget

Often you will be asked to customize look and feel of the chat widget for your business. In this section, you will learn how you can customize the look and feel of your chat widget quickly and efficiently.

In section three of "Add JavaScript Code," you have used the Bots.init() function. By simply passing some additional attributes to that, you can change various features of your widget. Check the following snippet of this function to achieve this:

```
<script>
  Bots.init(
    {
      appId: '5d20fd8bee36010010f274ae', // app id of channel
      menuItems: {  // enable/disable bottom-left corner options
        imageUpload: false,
```

```
      shareLocation: false,
      fileUpload: false,
    },
    browserStorage: 'sessionStorage', // when to clear chat history
    businessName: "I'm Travvy", // bold name text
    customText: {
      headerText: 'Book A Hiking Trip', // top pannel text
      inputPlaceholder: 'Start booking a trip...' //message input
                                                place holder
    },
    buttonIconUrl: './assets/Kompas kaart-02.png', // chat-widget
                                                button icon
    businessIconUrl: './assets/Kompas kaart-02.png', // widget
                                                response icon
    backgroundImageUrl: './assets/Travvy.png', // widget background
                                                image
  }
 )
</script>
```

In the code editor, this should look as Figure 6-23.

```
<script>
  Bots.init(
    {
      appId: '5d20fd8bee36010010f274ae', // app id of channel
      menuItems: {  // enable/disable bottom-left corner options
        imageUpload: false,
        shareLocation: false,
        fileUpload: false,
      },
      browserStorage: 'sessionStorage', // when to clear chat history
      businessName: "I'm Travvy", // bold name text
      customText: {
        headerText: 'Book A Hiking Trip', // top pannel text
        inputPlaceholder: 'Start booking a trip...' //message input place holder
      },
      buttonIconUrl: './assets/Kompas kaart-02.png', // chat-widget button icon
      businessIconUrl: './assets/Kompas kaart-02.png', // widget response icon
      backgroundImageUrl: './assets/Travvy.png', // widget background image
    }
  )
</script>
</body>
```

Figure 6-23. *Widget customization properties*

The following are a few additional default properties, in case you need to update:

1. To change the bot icon, search for the text in frame.1.16.1.min. js "**m.default.createElement("img",{alt:name+"'s avatar",src:t}))}}])**" and update it as "**m.default.createElement("img",{alt:name+"'s avatar",src:'http://localhost:8080/ odaTravvyDemo/Kompas.png'}))}}])**". Highlighted section is the location of the image.

2. To update the introductionText of the widget, search for "**introductionText:"We're here to talk, so ask us anything!""** and update it as "**introductionText:"Extreme Hiking Assistant!".**"

We would also advise you to refer to the README.md file for more customization options.

After accommodating all these changes, your widget would look as Figure 6-24.

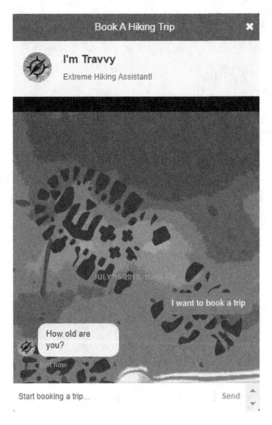

Figure 6-24. *Customized chat widget*

Summary

In this chapter, you were familiarized with various types of channels Oracle provides you to expose your Skills and Digital Assistants. You focused on the Users Web channel as this was the primary focus of the sample app which is being created for this book. Having said that, you are encouraged to explore various channel types mentioned in the initial sections of this chapter. Refer to Oracle documentations regarding the instructions for those channel types. Again, Agent Integration and Webhook channels will be explained in another chapter of this book. In the later sections of this chapter, you learned how, with bare minimum changes, you can radially enhance the look and feel of the chat widget. We hope that this chapter provided you enough information to start working with channels and will see you in the next chapter.

CHAPTER 7

Supporting Multiple Languages in Your Digital Assistant

Introduction

आपने शायद इस अध्याय को हिंदी के वाक्य के साथ प्रारम्भ करने का विचार नहीं किया होगा.

Or, in English: You might not have considered to begin this chapter with a Hindi sentence.

While this book is written in English, there are many more languages around the world, and many of these are spoken by only a minority of the world population. There are roughly 6,500 spoken languages in the world today. However, about 2,000 of those languages have fewer than 1,000 speakers. The ten languages spoken by most people cover around 50% of the world. This means that the other 50% speaks any of the remaining 6,490 languages. These figures pose a challenge to anyone wanting to provide a multilingual solution. It is key to determine what languages you want to support, as it is virtually impossible to support all.

A simple example says it all:

- I would like to order a pizza.

- Je voudrais commander une pizza.

- Me gustaria pedir una pizza.

- Ich mochte gerne einen Pizza bestellen.

How would one build a digital assistant that understands all the preceding languages in a way that they relate to the OrderPizza intent and also reply to the user in the correct language?

© Luc Bors, Ardhendu Samajdwer, Mascha van Oosterhout 2020
L. Bors et al., *Oracle Digital Assistant*, https://doi.org/10.1007/978-1-4842-5422-6_7

Extreme Hiking Holidays has the same challenge as they serve customers from multiple countries. Most of their customers speak English, but as an extra service to customers, they also provide their customers with a Digital Assistant that does not only speak English but also other languages. In this chapter, you will learn how to add support for **multiple languages** to Oracle Digital Assistant.

Note   Multilanguage support on a Digital Assistant is not perfect yet. For individual skills, you can have multilanguage support. If you need this multilanguage support, you should consider exposing skills. On the Digital Assistant level, there is currently no way to detect languages or to use resource bundles. The current solution for this is to create so-called "predominant" language bots. These are not multilanguage but single–foreign language bots.

Enabling Oracle Digital Assistant for true multilanguage use is on the roadmap.

How Do Translations Work?

Oracle Digital Assistants' NLP is based on English. This does not mean that you cannot add support for other languages. This support depends on two components.

First is the use of an external translation service. This will take any utterance that is entered by the user and translate it to and from English. Second is the use of resource bundles that contain the language-specific texts. These will be used to send translated bot responses to the user. Later in this chapter, we explain how these components work and how they can be combined. First, you will learn about a typical flow for language translation as displayed in Figure 7-1.

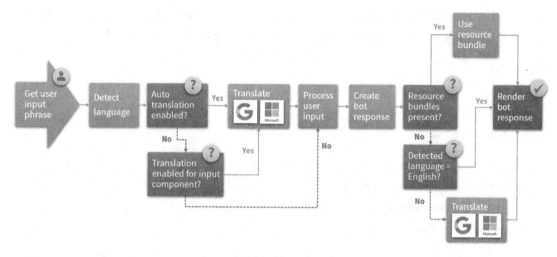

Figure 7-1. *The translation flow*

Based on the user input, the Digital Assistant will try to detect the language. Once the language is detected, it will proceed to the next step in the flow. In that step, if needed, the users' utterance will be translated to English by the translation service. This need for translation is determined by the **autoTranslate** property. The autotranslation is disabled by default, but at any time, it can be enabled or disabled. The **autoTranslate** context variable needs to be defined, and whenever the value of this variable is set to true, autotranslation is enabled. When enabled, all user messages are translated to English, and all the bot messages are translated to user language.

```
variables:
    autoTranslate: "boolean"

states:
    enableAutotranslation:
    component: "System.SetVariable"
    properties:
        variable: "autoTranslate"
        value: true
    transitions: {}
```

If **autoTranslate** is false, translation might still be required for individual components. This is due to the fact that the **translation** property on components can override the global **autoTranslate** setting.

So if the **translation** property on a component is set to **true**, the Digital Assistant will invoke the configured translation service, even if autoTranslate is set to false.

```
<aState>:
  component: "System.Text"
  properties:
    translate: true
```

After invoking the translation service, or when no translation is needed, the Digital Assistant will continue to process the users' input and, next, create the bots' response.

The Digital Assistant needs to figure out how the response is sent to the end user. There are three options:

1) This can be derived from a **resource bundle**.

2) This can be translated via a **translation service**.

3) This can be sent "**as is.**"

The exact way depends. By default, the output string defined on the component is used. If the defined output string is a reference to a resource bundle, then the resource bundle string is used. If the language setting is not English, then the resource bundle entry for this language is looked up and used. If it is not available, then the fall back is to the default language (English) bundle string. If translation is enabled for a component (which can be through autoTranslate or the translate property), then the component output text is sent to the translation service for translation first. This is the case even if the string is obtained from a resource bundle

---

Note   The internal language used in Oracle Digital Assistant is English. If the users' language is English, there is no need to translate the bots' reply, as both are English. In this case, you would need to use Apache FreeMarker expressions to set the translate property on the components to false or the autoTranslate property to false. Otherwise, Oracle Digital Assistant will try and translate to English.

If the user's language is other than English, Digital Assistant will invoke the translation service and communicate the translated answer to the user. However, if you set translate property to false on components that reference resource bundles, then the resource bundle string is used.

Note   If you start your flow with a System.TranslateInput, the Digital Assistant will ALWAYS translate the users input to English. This can be very helpful if you need this kind of translation.

```
translateInput:
    component: "System.TranslateInput"
    properties:
        variable: "translatedString"
    transitions:
        next: "Intent"
```

Hint   You can even read the original string from a variable, in which case you would keep copies of the original and the translated string.

Pitfalls in Translation Services

Translation services can be very helpful in creating multilingual Digital Assistants. However, they are not failsafe. There are several situations where translation services will have difficulties detecting the correct language or providing the correct translations. Let's have a look at some examples.

First there is the case of closely related languages. In Europe there are several languages that are very closely related. For a translation service, it will be difficult to distinguish between, for instance, Norwegian and Swedish and Swedish and Danish. In such case, you as a developer should be prepared and ask the user to confirm the detected language before proceeding. Second there are greetings or words that are "borrowed" from other languages. Hi, Hey, Hoi, Hello, Hallo, Bonjour, Ciao, all kinds of greetings that are used in many different languages. There is no translation service that is able to distinguish between an Italian saying "Ciao" and a Dutchman saying "Ciao." It will always come up with Italian as language, thus leaving the Dutchman in the dark. There is nothing you can do about this.

Adding Multilanguage Support to Travvy

Extreme Hiking Holidays has customers in over 50 countries, but they don't have employees that can cover all of these 50 languages. Besides that, they don't have the resources to create and manage a bot in 50 different languages. This is where the translation services come in handy.

Working with translation services requires the setup of an external translation service. Oracle Digital Assistant supports the use of the Google Translation API and Microsoft translation services.

Note   For the purpose of this book, we will use the Google Translation API.

Google Translation API is a paid API. Costs can vary and are based on API usage. Before you can use the API, you must set up a Google Account and enable the Translation API. The API can be enabled in the developer console (Figure 7-2) for your Google Account: `https://console.developers.google.com/apis/dashboard`.

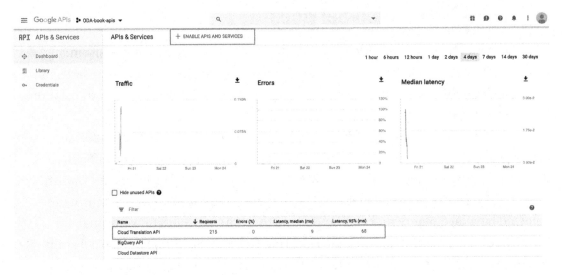

Figure 7-2. *Google developer console dashboard*

Once you have activated the translation service, this translation service has to be added to the Digital Assistant in order to use this. You can open the "Add Translation Services" from the "Settings menu" at the level of the Digital Assistant (Figure 7-3).

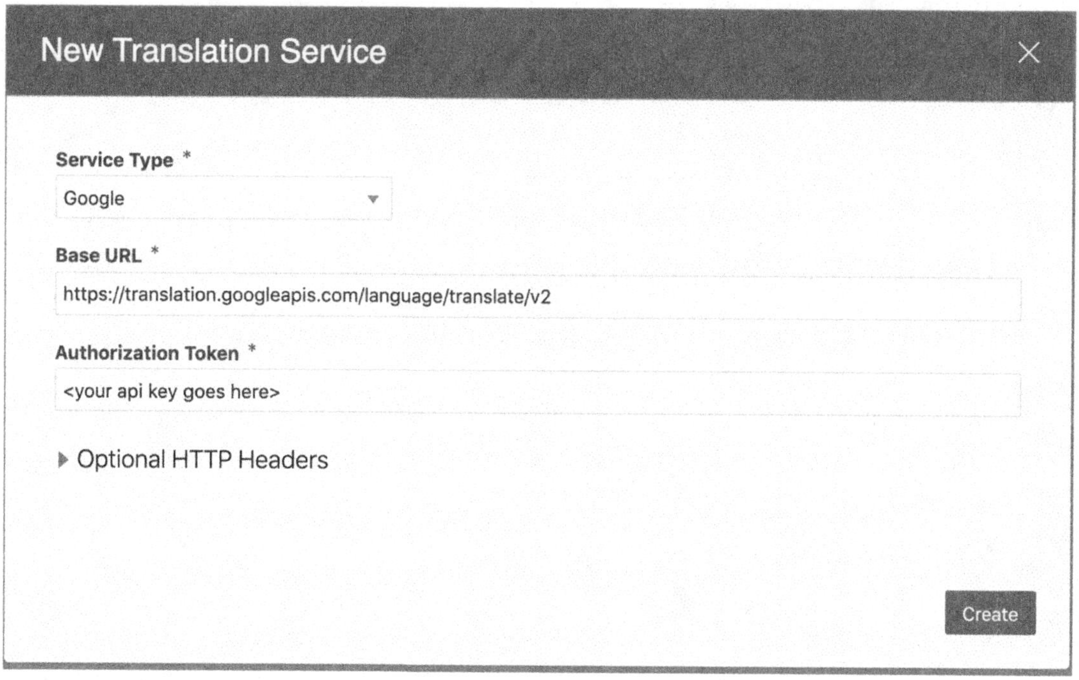

New Translation Service

Service Type *

Google

Base URL *

https://translation.googleapis.com/language/translate/v2

Authorization Token *

<your api key goes here>

▶ Optional HTTP Headers

Create

Figure 7-3. *Adding a translation service*

You simply select the service type from the dropdown, add the base URL for the translation service, and finally enter the Authorization Token. When you click create, the configuration is saved, and the translation service is ready to be used.

After adding the translation service, you need to configure its use. The usage of the translation service can be configured at skill level. In the skills' general settings, you need to select the translation service that you want to use from a dropdown (Figure 7-4) that contains all the available translation services for this cloud instance.

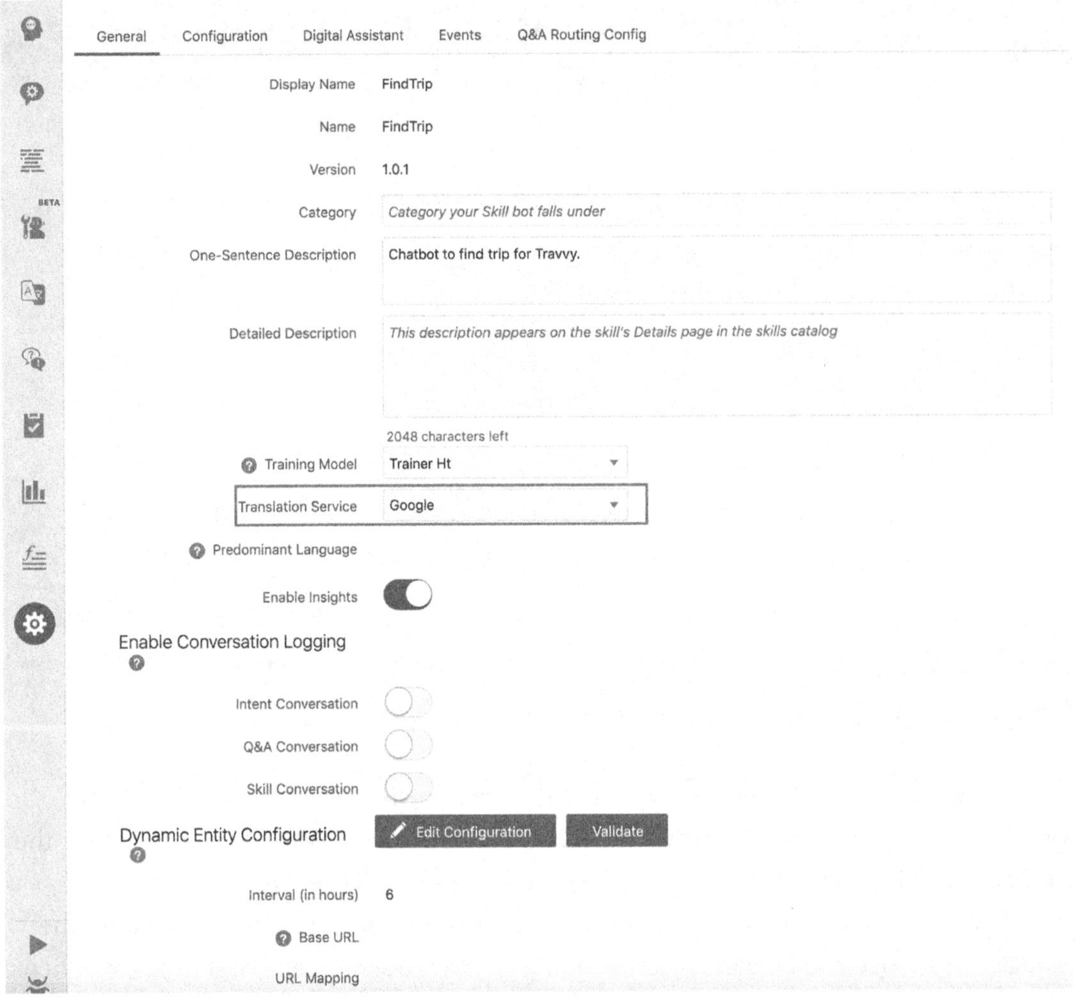

Figure 7-4. *Setting the translation service*

Once this is selected, all the needed translations will be provided by this translation service.

Using the Translation Service

For translations in Digital Assistant, there are two options, Opt-in and Opt-out. Both will be explained in this section.

Opt-out

To have translations in Digital Assistant enabled by default, also known as Opt-out
translation, there are some steps to take after adding the translation service. First you
need to add the **autoTranslate** context variable. Next you need a **System.SetVariable**
component that sets the **autoTranslate** to **true**. Finally, as the next state, you need to
add the **System.DetectLanguage** component, which calls the translation service to
determine the users' language. All of these will result in the following piece of code:

```
metadata:
  platformVersion: "1.0"
main: true
name: FindTripWithTranslations
context:
  variables:
    iResult: "nlpresult"
    autoTranslate: "boolean"

states:
  enableAutoTranslation :
    component: "System.SetVariable"
    properties:
      variable: "autoTranslate"
      value: true
    transitions:
      next: "detectLanguage"

  detectLanguage:
    component: "System.DetectLanguage"
    properties:
      source:
    transitions:
      next: ............
    ............
    ............
    ............
```

Opt-in

The other option for translations is the Opt-in option. Only for those components that you want to have translated, you specifically set the translate property to **true.**

In order to make this work, you must make sure to set the **autoTranslate** context variable to **false**. Next, for all individual components that need translation, set the **translate** property to **true**. Finally make sure that the **System.Intent** component also has the **translate** property set to **true**. This is because the internal language of the Oracle Digital Assistant is English.

```
intent:
  component: "System.Intent"
  properties:
    variable: "iResult"
    translate: true
```

Tip   If you rely on a translation service for the answers, the Translation API will translate the Bots' English answer to the users' language. You have little or no control on the translation, and thus, you are not 100% sure that the answers are always the same. To be in full control of what your Digital Assistant replies, you should use resource bundles to store the language dependent answers.

Using Resource Bundles

As stated in the previous tip, it is good practice to rely on resource bundles for all replies from the bot. This can be done for as many languages and dialects as you need. To implement this, you must create resource bundles and resource bundle entries. When you have not added any resource bundles yet, Oracle Digital Assistant will create one for you, and you can enter the first entry (Figure 7-5).

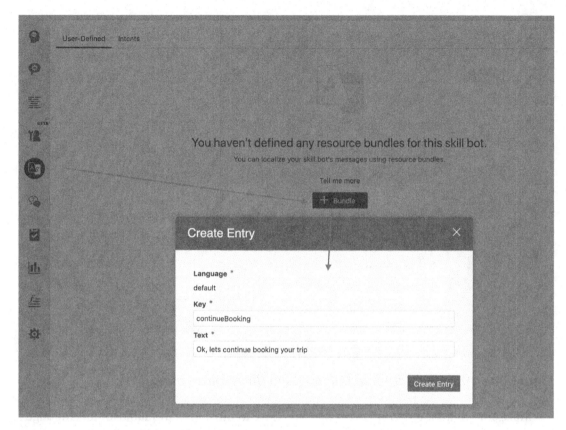

Figure 7-5. *Creating the first resource bundle entry*

Note   When you create a new resource bundle entry, it will always be added to the default language, which is English.

Now you can start adding translations, or in other words resource bundles for other languages. For every key, you can add a translation by invoking the "+ Language" button. This will show you a popup where you can enter language-specific text for the selected key (Figure 7-6).

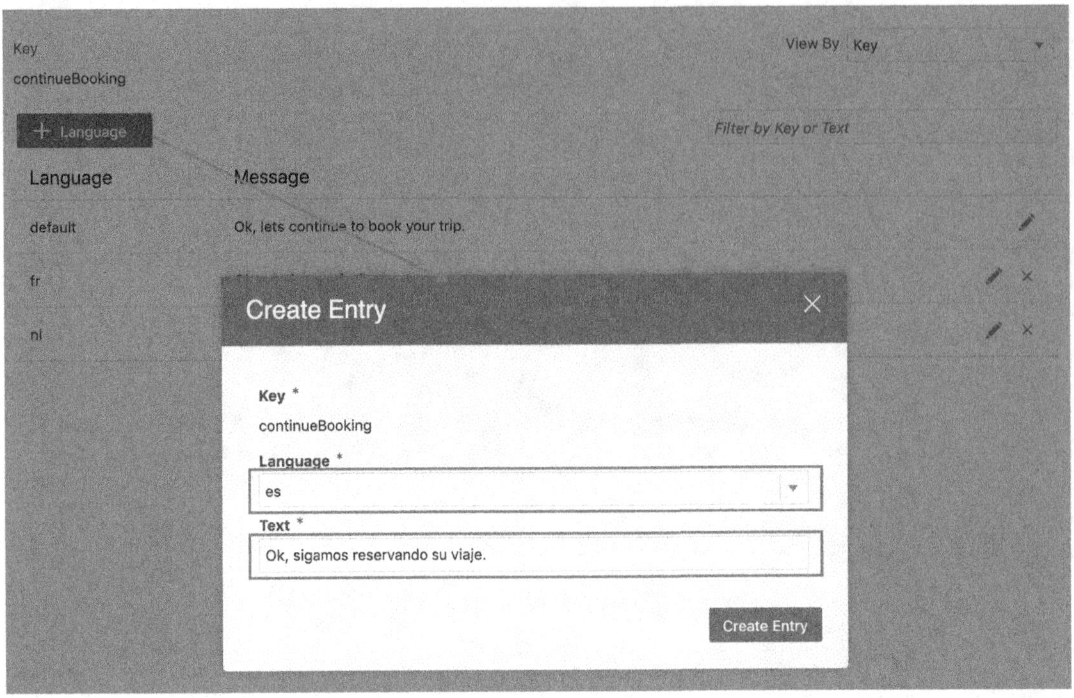

Figure 7-6. *Adding additional languages to the resource bundle*

Make sure that for each and every key, you have a translation available for all languages that the Digital Assistant supports. To help you with this, Oracle Digital Assistant has a "View By" dropdown at the top right (Figure 7-6). First, you can switch this to "language." Then you select the language to create resource strings for and start building strings for as long you see keys. When all keys have a string, then no more keys are shown.

Answering from a Resource Bundle

With all the translations entered, it is now time to make the Digital Assistant reply to the user based on these translations. This is actually very simple. A single-language Digital Assistant can use hardcoded text as shown in the following code sample:

```
bookTrip:
  component: "System.Output"
  properties:
    text: "Ok, lets continue to book your trip."
  transitions:
    return: "done"
```

To support multiple languages in a skill, you should add a reference to a resource bundle, so it can reply based on this resource bundle. The previous code sample will have to change to

```
bookTrip:
  component: "System.Output"
  properties:
    text: "${rb.continueBooking}"
  transitions:
    return: "done"
```

where **rb.continueBooking** refers to a resource bundle entry in the conversation's language having a key with value "continueBooking." The Digital Assistant will look this up and reply the exact value belonging to that key.

Note that the prefix **rb** needs to be defined in the context/variables section of the flow.

```
context:
  variables:
    iResult: "nlpresult"
    rb: "resourcebundle"
    autoTranslate: "boolean"
```

Note   You can make response more dynamic to use parameters in resource bundles. The following example shows how you could add dynamic pricing for hotel rooms. The bot replies with text from a resource bundle which dynamically changes the price based on the outcome of a service call in a custom component.

The resource bundle key and value:

Key: Pricing.

Value: Currently the price for this hotel is {0} per night.

The oBOTMLcode:

```
component: "System.Output"
    properties:
text: ${rb(pricing, '${pricingService.value}')}"
    transitions:
      return: "done"
```

Determine Language from Utterance

In the previous section, you learned how to use the language from the user profile in the conversation and have everything translated from and to that language. There is another option, and that is to use the translation service to determine the users' language. In order to make this happen, we need to use "**System.DetectLanguage**". This component uses the translation service to detect the user's language and automatically sets a variable named **profile.languageTag** with the locale string.

Note   The **profile.languageTag** takes precedence over the **profile.locale** variable.

```
metadata:
  platformVersion: "1.0"
main: true
name: FindTripWithTranslations
context:
  variables:
    iResult: "nlpresult"
    rb: "resourcebundle"
    autoTranslate: "boolean"

states:

  enableAutoTranslation:
    component: "System.SetVariable"
    properties:
      variable: "autoTranslate"
      value: false
```

```
transitions:
  next: " detectLanguage"

detectLanguage:
  component: "System.DetectLanguage"
  properties:
    source:
  transitions:
    next: "intent"

intent:
  component: "System.Intent"
  properties:
    variable: "iResult"
  transitions:
    actions:
      SelectTrip: "askAge"
      CancelTrip: "cancelTrip"
      unresolvedIntent: "unresolved"
```

If you enter an utterance in any language that is supported by your digital assistant, it will resolve the intent and answer the user in the same language. As illustrated in Figure 7-7, the Digital Assistant recognized that the user's utterance is English, and it replies in English to the user. Next, the user's utterance is in Dutch, it is clearly understood by the Digital Assistant, and it switches to the new language and replies in Dutch.

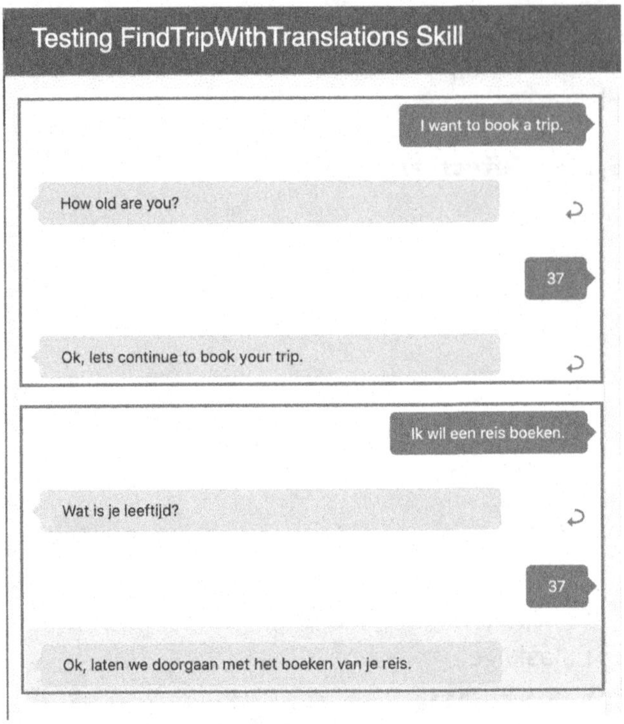

Figure 7-7. Digital Assistant answering in users' language

There always is the possibility that the users' language is not supported by your Digital Assistant.

Note   The restriction is with what the translation service supports. Good design practice is that bot designers only support those languages that make sense to their business and that they have language skills in-house.

There are several ways of communicating this to the user. A very nice and user-friendly way is to tell the user in their own language that their language is not supported. This message can be accompanied with a list of supported languages. The user can select one of the supported languages, and you can store the selected language in the **profile.languageTag**.

In order for this to work, an additional piece of logic needs to be added to the flow, right between the **detectLanguage** state and the **intent** state.

The next code sample shows how to implement this functionality:

```
detectLanguage:
  component: "System.DetectLanguage"
  properties:
    source:
  transitions:
    next: "checkUserDetectedLanguage"

checkUserDetectedLanguage:
  component: "System.Switch"
  properties:
    source: "${['nl','fr','en']?seq_contains(profile.languageTag)?string
    ('yes','no')}"
    values:
    - "yes"
    - "no"
  transitions:
    actions:
      yes: "Intent"
      no: "selectLanguageFromList"
      NONE: "selectLanguageFromList"

selectLanguageFromList:
  component: "System.CommonResponse"
  properties:
    processUserMessage: true
    keepTurn: false
    translate: true
    metadata:
      responseItems:
      - type: "text"
        text: "${rb.globalLanguagePrompt}"
        actions:
        - label: "${rb.globalLanguageEnglish}"
          type: "postback"
          payload:
```

```
              variables:
                profile.languageTag: "en"
        - label: "${rb.globalLanguageDutch}"
          type: "postback"
          payload:
            variables:
              profile.languageTag: "nl"
        - label: "${rb.globalLanguageFrench}"
          type: "postback"
          payload:
            variables:
              profile.languageTag: "fr"
  transitions:
    next: "Intent"
```

With this in place, whenever Travvy does not support the user's language, it will reply in that language that Travvy only supports – English, French, Dutch, and Spanish. Figure 7-8 shows how that would look in Italian.

Figure 7-8. Reply to the user in their own language

Note   While you can use language detection, you must realize that this has an effect on performance. With language detection "on," for each and every utterance, the translation service will be called to detect the language, thus adding calls to external service to the conversation.

Using Language from the User Profile

As stated in the previous section, using a translation service for language detection can have a small negative performance effect. There is another approach that takes away this effect by omitting the language detection. This approach uses the language from the user profile. So instead of the automatic setting of the **profile.languageTag** by the translation service, you now set it explicitly as in the following code sample:

```
metadata:
  platformVersion: "1.0"
main: true
name: FindTripWithTranslations
context:
  variables:
    iResult: "nlpresult"
    rb: "resourcebundle"
    autoTranslate: "boolean"

states:
  enableAutoTranslation:
    component: "System.SetVariable"
    properties:
      variable: "autoTranslate"
      value: false
    transitions:
      next: "setBrowserBasedProfileLanguage"

#Set profile.languageTag based on user's browser language
```

```
setBrowserBasedProfileLanguage:
  component: "System.SetVariable"
  properties:
    variable: "profile.languageTag"
    value: "${profile.locale}"
  transitions:
    next: "Intent"
```

In order for this to work, you need to get the users' locale. One way to achieve this is to read the **profile.locale** variable which is set from the messaging client.

Another way to do this is to explicitly set it on the messaging client. The following code sample shows how this can be done in JavaScript using the Oracle Digital Assistant JavaScript SDK (see Chapter 6) when you invoke the Digital Assistant from a web page:

```
Bots.updateUser(
  {
  "properties": {
    "userLocale": userLocale}
  }
);
```

Summary

There are many languages spoken around the world. In order for Digital Assistants to be successful, they should be able to speak other languages than just English. Luckily this requirement of multilanguage support can be implemented in Oracle Digital Assistant. In this chapter, you learned how to do that based on translation services and resource bundles. It is not 100% guaranteed that translation services cover all use cases; neither will they be able to cover all utterances. You will have to help the user and the translation service by providing functionality to either pick or confirm the language the user prefers.

CHAPTER 8

Using Webview Components

Chatbots are the conversational approach of retrieving and providing information to end users. An end user enters his/her query to the bot, and the bot responds back with appropriate answers. One of the core reasons of chatbots being so popular in today's world is that they can provide response in much faster way with bare minimum user inputs.

As the popularity of chatbots is increasing day by day, organizations are using chatbots not only to provide information about their offerings, but also they are using them for business-critical flows such as ordering, booking, providing feedbacks, and so on. Often such flows in chatbots require multiple user information which an end user needs to provide in order to complete certain transaction/task. Such a long chain of user interaction with a bot might become a bit boring from an end user point of view. As a result, the user may rather prefer to use a different route to fulfill his/her requirement such as calling customer care or going back to the organization's web site.

Oracle acknowledged this fact and came up with the concept of Webview components for Digital Assistants. Using Webview components, parameters can be passed from the skill to the web form and from the web form back to the skill. In this chapter, you will become familiar with the Webview components and their uses.

What Is a Webview Component?

Though chatbots are designed for a conversational approach of information gathering and sharing, it is not always ideal. An alternate approach to handle such cases can be providing a web form to the end user, which he/she fill at once in order to complete a specific task. You can craft your bot flow in such a way that it provides such a form to

179

© Luc Bors, Ardhendu Samajdwer, Mascha van Oosterhout 2020
L. Bors et al., *Oracle Digital Assistant*, https://doi.org/10.1007/978-1-4842-5422-6_8

the end user, only for those scenarios which require structural data input, for example, providing a web form containing necessary address fields while placing an order or a form with fields which are essential to complete a holiday booking or fields with mandatory information required for opening a bank account and so on.

The System.Webview component, as the name suggests, allows you to integrate a web page with your digital assistant in order to accumulate structural data from end users. Though Webview components are integrated with the digital assistant, they open in a separate browser tab/window for web-based chat clients or in webview in case of mobile devices outside of the chat client. Because of this, you are able to leverage various features like form view, select choices, date pickers, custom styles, and so on in the web page. But it doesn't end here. In addition to such features, you can also add various kinds of input validations on your web page with ease, such as mandatory fields, email format, date range, and so on. This validation logic stays on the web page, and you don't need to add them inside your digital assistant.

You are already aware that components are broadly classified into two categories in the context of the Oracle Digital Assistant, namely, built-in components and Custom Components. Webview components (System.Webview) fall under the category built-in components, which means they are provided by Oracle out of the box.

Using a Webview component in your dialog flow, you can route your end user to a web application where he/she can provide the necessary inputs to complete a transaction. Since Webview components are provided by Oracle out of the box, you can easily add them to your flow by clicking the **+ Components** button on the Flows screen and then selecting the same from the **User Interface** component.

Refer to Figure 8-1.

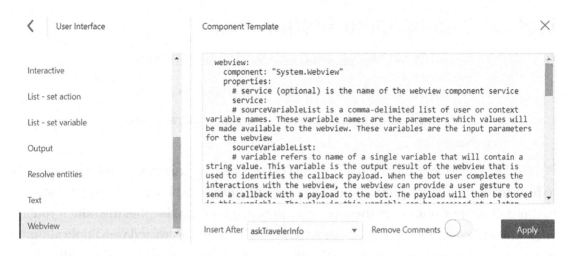

Figure 8-1. *Webview component*

Before we start discussing Webview component further, let's have a look at the pros and cons.

Pros:

- Use of a web application to gather user information instead of an extensive bot communication.

- Web application screen size will not be limited to chat client window.

- Web application can be custom skinned as preferred.

- Prebuild web application can also be used (web page should be accessible using GET request).

- Leverage various browser features such as upload and download.

- Data entered by user remain secure as it doesn't appear in the chat session.

Cons:

- As Webview opens the application in a separate browser window, the application doesn't look integrated with the chatbot. And hence, the user might get confused if not preinformed properly in the chat before application launch.

Webview Component Architecture

You can adhere to two different approaches in order to implement Webview components:

- Webview application hosted within Oracle Digital Assistant

- Webview application hosted externally (outside of Oracle Digital Assistant)

Similar to Custom Components, Oracle Digital Assistant provides a built-in container to host single-page web applications. If you prefer to locally deploy your web application in this built-in container, then the Webview application is hosted within the skill. If you wish to use your existing web application or deploy your newly created web application in a remote environment, outside of Digital Assistant, then that will be an externally hosted Webview application. In both cases, the application is launched in a separate browser tab for web-based chat clients or in webview in case of mobile devices. Let's take a deeper look in both of these approaches.

Webview Application Hosted Within Digital Assistant

Applications deployed locally in your Oracle Digital Assistant should be single-page applications (SPAs). In case you are unfamiliar, SPA renders a single page and has a single entry point for the application, but page content can be changed/updated dynamically at runtime. You can use any JavaScript framework to develop such an application for the System.Webview component, for example, React, Oracle Jet, Oracle VBCS, and so on. You then bundle your application in a TAR archive (with a .tgz extension) and deploy it into the local container to create a Webview service.

Note   The entry point of your SPA must be an index.html file, placed at the root level of your application and, hence, your deployment archive.

When you use a System.Webview component in your dialog flow, it then

1. Invokes the index.html for you SPA and launches the app in a new browser tab.

2. System.Webview component then passes input parameter(s) from your dialog flow to your SPA along with a callback URL. The callback URL is generated and appended to the webview request by the System.Webview component as **window. webViewParameters** parameter.

3. Your SPA makes a POST request to the callback URL upon completion of its processing. The callback URL can be accessed from the parameter **webview.onDone** in the app. In case you need to pass any data back to the Webview component in the skill from your app, then that data is passed as **JSON object**, which gets stored in the dialog flow variable that is referenced from the System.Webview variable property. You can easily access this data in your dialog flow as **${variable _name.value.Param}**.

A pictorial representation of this flow can be found in Figure 8-2.

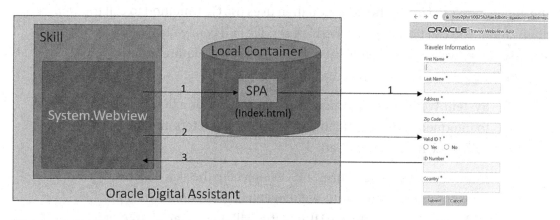

Figure 8-2. *Local deployment of web application*

The abovementioned flow will be explained in detail later in this chapter, under section "Webview Application Hosted Within Digital Assistant".

Note   If you wish to use Oracle VBCS to create your SPA, it will have some limitations. Since the backend of VBCS is responsible for business objects, REST calls made to the business objects, as well as the user access for the business objects (using Oracle Identity Cloud Service, IDCS), these features will not be available when you export your application. If you wish to use Oracle VBCS in order to build your SPA, then you must deploy the app externally and take care of these features explicitly.

Webview Application Hosted Externally

If you already have an existing application which you want to use as Webview, then you need to take this approach. These are typical cases when your application has security requirements or you need to use server-side infrastructures and so on. Unlike the case of a locally deployed application, your external application may or may not be a SPA. Only requirement is that the application, which you are willing to use, should be accessible over a GET request from the skill. In the case of an externally hosted web application, you need to use an intermediary service which will be responsible for composing the application call URL to the skill. Skill then uses the GET method to call the web application. You can host the intermediary service and your web application on the same server as well as on different servers based on your requirement.

The flow goes as mentioned in the following:

1. **System.Webview** component from the skill sends a POST request to intermediary service along with input parameters and the callback URL in the payload.

2. Intermediary service returns a JSON object to the skill which contains only one property, **webview.url,** which holds web app URL as well as the callback URL to the skill. In case any parameter needs to be passed to the web app, then those parameters get appended to webview.url property. System.Webview component in the skill then uses this property to make all subsequent GET requests to the web app.

3. Skill sends a GET request to launch the app.

4. The web application processes the request and makes a POST call to the callback URL indicating that the app has finished its processing. Again, in case you need to pass any data back to the Webview component in your skill from the web app, then that data is passed as **JSON object**, which gets stored in the dialog flow variable that is referenced from the System.Webview variable property. You can then access this data in your dialog flow as **${variable _name.value.param_name}**.

Figure 8-3 shows a pictorial representation of the preceding flow with the intermediary service and web application hosted on the same web server.

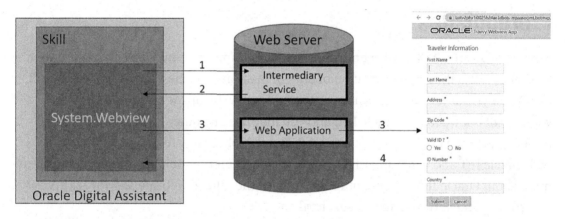

Figure 8-3. *Externally hosted web application*

Figure 8-4 shows the flow when the intermediary service and the web application reside on separate servers.

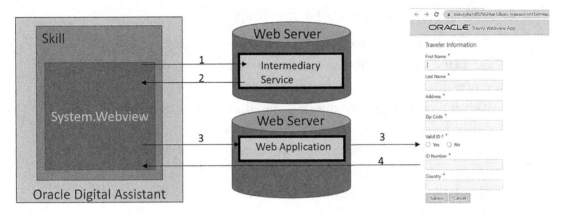

Figure 8-4. *Externally hosted web application with service and application on different servers*

Prerequisites

Now that you are familiar with both approaches of implementing Webview component, let's take look at the implementation of both of these approaches. In both cases, we will use Oracle JET as JavaScript framework to build our web app. As mentioned before, you are not bound to use Oracle JET, and you may use any other framework of your preference. In the upcoming sections, we will not explain how to design applications with Oracle JET and will primarily focus on the implementation part. If you want to learn more about Oracle JET, we encourage you to visit the official web site: `www.oracle.com/webfolder/technetwork/jet/index.html`.

The following is the list of prerequisites which you need to have/install in order to follow the implementations:

1. Access to a Digital Assistant environment (19.1.5 or higher).

2. Download and install Node.js (`https://nodejs.org/en/download/`).

3. Microsoft Visual Studio Code IDE (`https://code.visualstudio.com/download`).

4. Install Oracle JET Command Line Interface (post-Node.js installation).

    Windows

    `npm install -g @oracle/ojet-cli`

MAC

```
sudo npm install -g @oracle/ojet-cli
```

5. Tunneling software to expose your system over the Internet. In our
    case, we will be using ngrok (`https://ngrok.com/download`). This
    will be needed for a web application hosted externally.

Use Case

In upcoming sections, you will be implementing Webview component for Travvy. You
will create a web application to retrieve traveler information for booking Extreme Hiking
Holidays' package. Figure 8-5 shows the mockup of the web application.

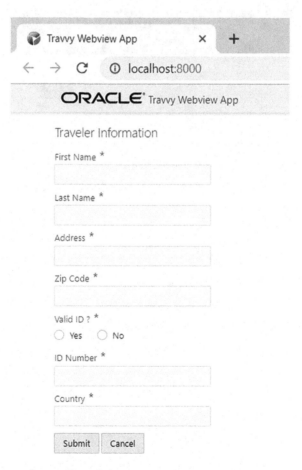

Figure 8-5. *Travvy web application*

Clicking the Submit button, user profile information will be processed, and the handle will be passed back to the skill, and the user will be notified accordingly. If user clicks Cancel button, then also we will pass the handle to the skill but will inform the skill that user clicked Cancel button. Accordingly, a message will be displayed back to the user. We are going to use Skill Tester for this demonstration, but you can also expose the skill over a channel as described in Chapter 6. Let's have a look at the flow as displayed in Figure 8-6.

1. User greets the skill with "Hi."

2. Skill replies back asking user to enter his/her first name.

3. User enters his first name.

4. Skill then gives a prompt asking the user to enter traveler's information to complete the booking.

Figure 8-6. *Execution of Webview component*

5. When the user clicks "Enter Traveler Information," the web app is launched in a separate window as you can see in Figure 8-7.

6. Application launch with prefilled first name as provide in the skill.

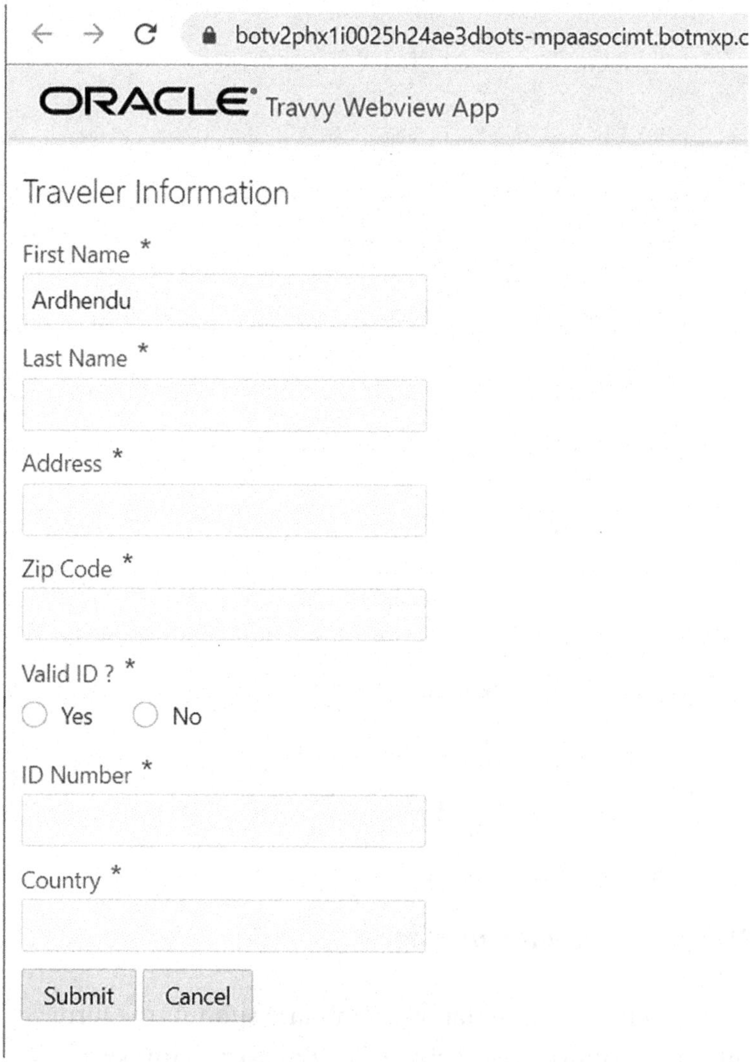

Figure 8-7. *Web app invoked from Skill with first name prepopulated*

7. If the user clicks the Submit button without entering the mandatory fields, validation messages are triggered as can be seen in Figure 8-8.

189

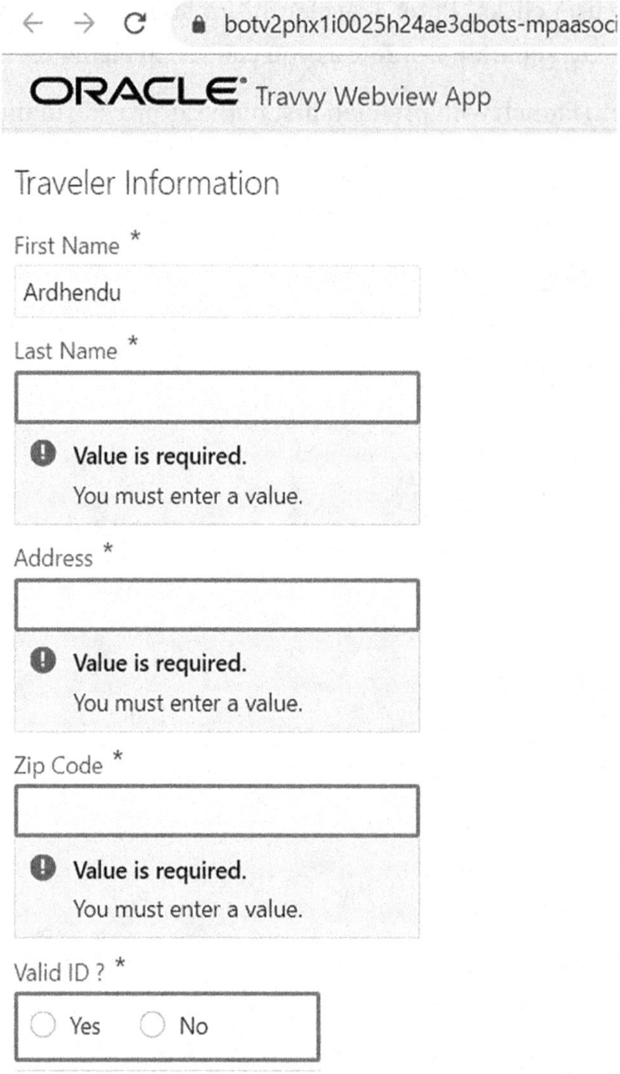

Figure 8-8. *Web app validation messages*

8. Once the user enters mandatory details and submits the form, a confirmation message is displayed to the user. In this case, user's full name is displayed based on web application's response (see Figure 8-9).

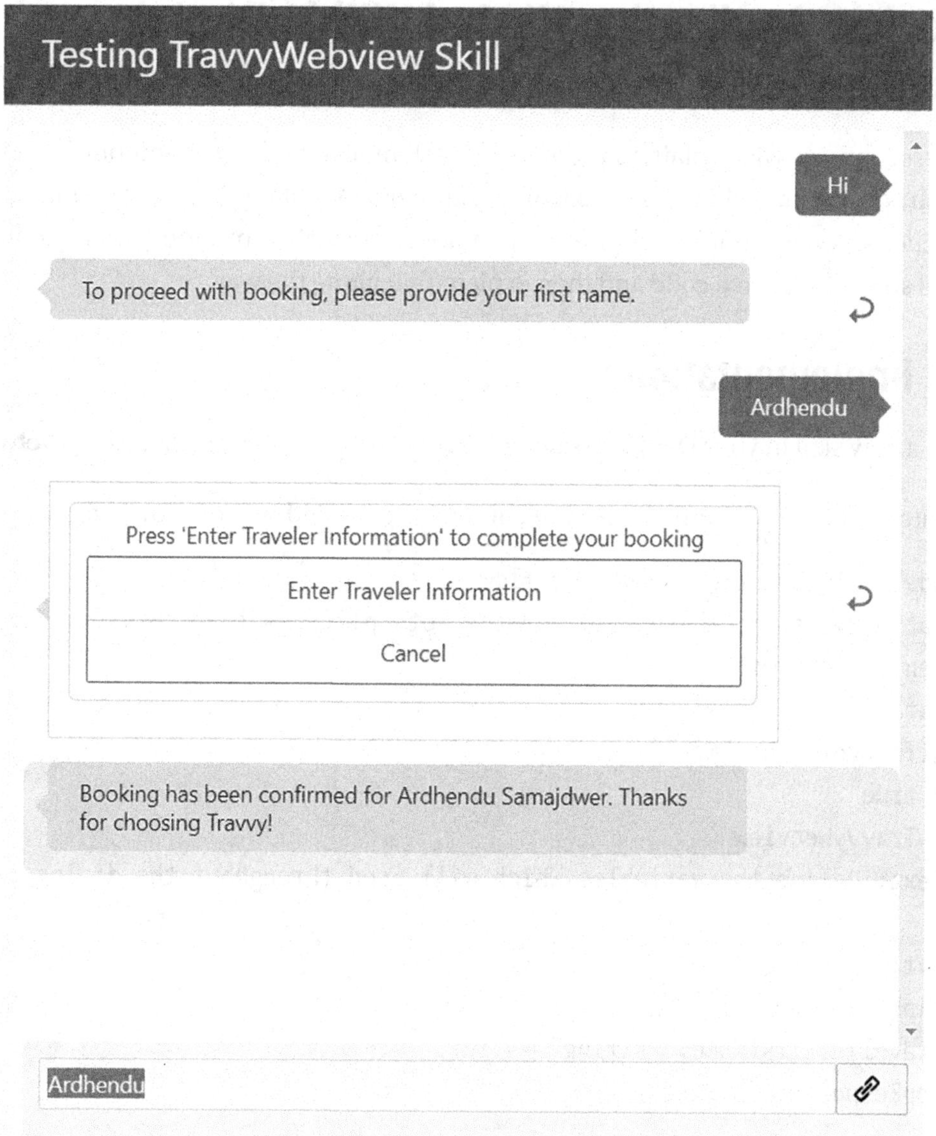

Figure 8-9. *Confirmation message with full name*

Webview Application Hosted Within Digital Assistant

In this section, we will explain to you how to implement a Webview component with the web application being deployed within the Digital Assistant. In the following subsections, you will find the skill implementation and web application implementation. We will share the source code and then explain that along the way.

Skill Implementation

Create a new skill in your Digital Assistant environment with the name **TravvyWebview**.

Navigate to Flows [icon] and replace the content as provided in the following:

```
#metadata: information about the flow
#  platformVersion: the version of the bots platform that this flow was
written to work with
metadata:
  platformVersion: "1.0"
main: true
name: TravvyWebview
#context: Define the variables which will used throughout the dialog flow
here.
context:
  variables:
    travellerFirstName: "string"
    appResponse: "string"

#states is where you can define the various states within your flow.

states:
  askTravelerInfo:
    component: "System.Text"
    properties:
      prompt: "To proceed with booking, please provide your first name."
      variable: "travellerFirstName"
    transitions:
      next: "webview"
```

```
webview:
  component: "System.Webview"
  properties:
    service: "travvyWebview"
    sourceVariableList: "travellerFirstName"
    variable: "appResponse"
    prompt: "Press 'Enter Traveler Information' to complete your booking"
    linkLabel: "Enter Traveler Information"
    cancelLabel: "Cancel"
  transitions:
    next: "evaluateWebviewResponse"
    actions:
      cancel: "onCancel"

evaluateWebviewResponse:
  component: "System.Switch"
  properties:
    source: "${appResponse.value.status}"
    values:
    - "success"
    - "cancel"
  transitions:
    actions:
      success: "confirmBooking"
      cancel: "onCancel"
      NONE: "onCancel"

confirmBooking:
  component: "System.Output"
  properties:
    text: "Booking has been confirmed for ${appResponse.value.firstName}
    ${appResponse.value.lastName}. Thanks for choosing Travvy!"
    keepTurn: false
  transitions:
    return: "done"
```

```
onCancel:
  component: "System.Output"
  properties:
    text: "Sorry that you canceled your booking. Hope to see you back soon!"
  transitions:
    return: "done"
```

1. At "context" level, you have defined two "variables":

    - travellerFirstName: Variable to store user's first name and pass it to the web application for booking

    - appResponse: Variable to store response from the web application

2. In the state **askTravelerInfo**, you asked user to provide first name. Notice that you are storing user's response in "travellerFirstName" variable.

3. In the next state **webview**, you added System.Webview component by clicking ➕ Components and then selected Webview from the User Interface component category. Take a look at the various properties defined for the component:

    - service: Name of the Webview service. You will create this service in the upcoming section.

    - sourceVariableList: List of source variable(s), which will be passed to the web application.

    - variable: Variable where the response of the web application will be stored.

    - prompt: Message prompt which will be displayed to the user asking them to provide additional information in the web application.

    - linkLabel: Label of the button, clicking which the web application will be launched.

    - cancelLabel: Label of the Cancel button, in case the user doesn't wish to provide user's information for booking.

    - transitions: Here you have defined the transition flow for the next state to evaluate application response and to cancel.

4. Next state in the dialog flow is **evaluateWebviewResponse**. In this state, you are evaluating the response of the web application, **${appResponse.value.status}**. In the web application, you will define two statuses, "success" and "cancel." In this state, you are checking whether the status from the web application is success or cancel. Based on the status, you then invoke the appropriate state.

5. In case of "success" status from web application, you invoke **confirmBooking** state. In this state, you are displaying confirmation message to the user based on the values entered in the web application. Check the expression you used for this purpose **${appResponse.value.firstName} ${appResponse. value.lastName}**.

6. In case of "cancel" status from web application or cancel action from the System.Webview component, you invoke **onCancel** state. In this state, you gracefully handled user's cancel response and displayed an appropriate message.

Web Application Implementation

Once you have Oracle JET CLI installed on your system, you can easily create a sample application by issuing the command

```
ojet create <project name> --template=navdrawer|navbar|basic|blank
```

in command prompt or on terminal from the location where you would like to create the application. For this demonstration, you can simply invoke the following command:

```
ojet create travvyWebviewApp --template=basic
```

1. Once done, open your appController.js (\travvyWebviewApp\src\ js\appController.js) and update the app name to "Travvy Web App" as follows:

    ```
    self.appName = ko.observable("Travvy Webview App");
    ```

2. Open index.html file (\travvyWebviewApp\src\index.html) and update **title** under **head** section as follows:

    ```
    <title>Travvy Webview App</title>
    ```

3. Replace the body section of your same index.html
    (\travvyWebviewApp\src\index.html) with the following code:

```
<body class="oj-web-applayout-body">
  <div id="globalBody" class="oj-web-applayout-page">
    <!--
        ** Oracle JET V7.0.1 web application header pattern.
        ** Please see the demos under Cookbook/Patterns/App
           Shell: Web
        ** and the CSS documentation under Support/API Docs/Non-
           Component Styling
        ** on the JET website for more information on how to use
           this pattern.
    -->
    <header role="banner" class="oj-web-applayout-header">
      <div class="oj-web-applayout-max-width oj-flex-bar oj-sm-
      align-items-center">
        <div class="oj-flex-bar-middle oj-sm-align-items-
        baseline">
          <span role="img" class="oj-icon demo-oracle-icon"
          title="Oracle Logo" alt="Oracle Logo"></span>
          <h1 class="oj-sm-only-hide oj-web-applayout-header-
          title" title="Application Name">
            <oj-bind-text value="[[appName]]"></oj-bind-text>
          </h1>
        </div>
      </div>
    </header>
    <div role="main" class="oj-web-applayout-max-width oj-web-
    applayout-content">
      <oj-validation-group id="tracker">
        <oj-form-layout id="form-container">
          <h3>Traveler Information</h3>
          <oj-input-text id="firstName" value="{{firstName}}"
          required="true" label-hint="First Name"></oj-input-text>
```

```
<oj-input-text id="lastName" value="{{lastName}}"
required="true" label-hint="Last Name"></oj-input-text>
<oj-input-text id="address" value="{{address}}"
required="true" label-hint="Address"></oj-input-text>
<oj-input-text id="zipCode" value="{{zipCode}}"
required="true" label-hint="Zip Code"></oj-input-text>
<oj-radioset id="validId" value="{{validId}}" class='oj-
choice-direction-row' required
  label-hint="Valid ID ?">
  <oj-option value="yes">Yes</oj-option>
  <oj-option value="no">No</oj-option>
</oj-radioset>
<oj-input-text id="idNumber" value="{{idNumber}}"
required="true" label-hint="ID Number"></oj-input-text>
<oj-input-text id="country" value="{{country}}"
required="true" label-hint="Country"></oj-input-text>
<div id='buttons-container'>
  <oj-button id='Submit' on-oj-action='[[buttonClick]]'>
  Submit</oj-button>
  <oj-button id='Cancel' on-oj-action='[[buttonClick]]'>
  Cancel</oj-button>
</div>
</oj-form-layout>
</oj-validation-group>
</div>
</div>

<script type="text/javascript" src="js/libs/require/require.
js"></script>
<script type="text/javascript" src="js/main.js"></script>

</body>
```

In this body section, you have defined various screen elements as displayed earlier in Figure 8-5 and added the on-screen validation.

4. Finally, open main.js (\travvyWebviewApp\src\js\main.js) file of
    your application, and replace the entire content of the file with the
    following code:

```
/**
 * @license
 * Copyright (c) 2014, 2019, Oracle and/or its affiliates.
 * The Universal Permissive License (UPL), Version 1.0
 */
'use strict';

/**
 * Example of Require.js boostrap javascript
 */

requirejs.config(
  {
    baseUrl: 'js',

    // Path mappings for the logical module names
    // Update the main-release-paths.json for release mode when
       updating the mappings
    paths:
    //injector:mainReleasePaths
    {
      'knockout': 'libs/knockout/knockout-3.5.0.debug',
      'jquery': 'libs/jquery/jquery-3.4.1',
      'jqueryui-amd': 'libs/jquery/jqueryui-amd-1.12.1',
      'promise': 'libs/es6-promise/es6-promise',
      'hammerjs': 'libs/hammer/hammer-2.0.8',
      'ojdnd': 'libs/dnd-polyfill/dnd-polyfill-1.0.0',
      'ojs': 'libs/oj/v7.0.1/debug',
      'ojL10n': 'libs/oj/v7.0.1/ojL10n',
      'ojtranslations': 'libs/oj/v7.0.1/resources',
      'text': 'libs/require/text',
      'signals': 'libs/js-signals/signals',
      'customElements': 'libs/webcomponents/custom-elements.min',
```

```
      'proj4': 'libs/proj4js/dist/proj4-src',
      'css': 'libs/require-css/css',
      'touchr': 'libs/touchr/touchr'
    }
    //endinjector
  }
);

/**
 * A top-level require call executed by the Application.
 * Although 'ojcore' and 'knockout' would be loaded in any case
   (they are specified as dependencies
 * by the modules themselves), we are listing them explicitly to
   get the references to the 'oj' and 'ko'
 * objects in the callback
 */
require(['ojs/ojcore', 'knockout', 'appController', 'jquery',
'ojs/ojknockout', 'ojs/ojformlayout', 'ojs/ojinputtext',
'ojs/ojlabel', 'ojs/ojbutton', 'ojs/ojradioset', 'ojs/
ojvalidationgroup'],
  function (oj, ko, app, $) { // this callback gets executed when
  all required modules are loaded
    $(function () {
      /*
        Parameters send from the webview are saved in a window
        object named "webviewParameters". In this sample
        the parameters contain information provided by the user in
        the bot conversation.
      */

      let webviewParameters = window.webviewParameters != null ?
      window.webviewParameters['parameters'] : null;

      /*
        helper function to read named webview parameter
      */
```

```javascript
let getWebviewParam = (arrParams, key, defaultValue) => {
  if (arrParams) {
    let param = arrParams.find(e => {
      return e.key === key;
    });
    return param ? param.value : defaultValue;
  }
  return defaultValue;
};
/*
  Setting default values for the travellerFirstName if no
  value provided
*/
self.firstName = ko.observable(getWebviewParam(webviewParame
ters, 'travellerFirstName', "));
self.lastName = ko.observable();
self.address = ko.observable();
self.zipCode = ko.observable();
self.validId = ko.observable();
self.idNumber = ko.observable();
self.country = ko.observable();

/*
  When the user submits ot cancels the web form, control
  need to be passed back to the bot,
  For this a callback URL is passed from the webview to the
  web application. The parameter
  holding the information is "webview.onDone"
*/

var webViewCallback = getWebviewParam(webviewParameters,
'webview.onDone', null);

this.tracker = ko.observable();
self.buttonClick = function (event) {
  let userProfile = {};
  userProfile.firstName = self.firstName();
```

```javascript
userProfile.lastName = self.lastName();
userProfile.address = self.address();
userProfile.zipCode = self.zipCode();
userProfile.validId = self.validId();
userProfile.idNumber = self.idNumber();
userProfile.country = self.country();

var tracker = document.getElementById("tracker");

if (tracker.valid === "valid") {
  // submit the form
  if (event.currentTarget.id === 'Submit') {
    userProfile.status = "success"
    //JQuery post call
    $.post(webViewCallback, JSON.stringify(userProfile));
  }
}
else if (event.currentTarget.id === 'Cancel') {
  let userProfile = {};
  userProfile.status = "cancel"
  //JQuery post call
  $.post(webViewCallback, JSON.stringify(userProfile));
}
else {
  // show messages on all the components
  tracker.showMessages();
  return;
}

const sleep = (milliseconds) => {
  return new Promise(resolve => setTimeout(resolve,
  milliseconds))
}
//Closes the browser tab the window is opened in. When
closing the browser
//tab, ensure the ajax call gets trough before the tab
closes. Thus adding
```

```
            //a "sleep" time
            sleep(500).then(() => {
              window.open(", '_self').close();
            })
            return true;
          }
          function init() {
            // Bind your ViewModel for the content of the whole page
            body.
            ko.applyBindings(app, document.getElementById('globalBody'));
          }
          // If running in a hybrid (e.g. Cordova) environment, we
          need to wait for the deviceready
          // event before executing any code that might interact with
          Cordova APIs or plugins.
          if ($(document.body).hasClass('oj-hybrid')) {
            document.addEventListener("deviceready", init);
          } else {
            init();
          }
        });
      }
    );
```

You can see the inline code comments in the preceding code specifying each and every section. Take a close look at **webviewParameters**, which is passed from the skill to your application and how that is being accessed using **getWebviewParam** inside the application. Check how you are setting "success" and "cancel" status and making a Ajax POST request using JQuery to the callback URL. **webview.onDone** parameter holds this callback URL.

Once everything is done, test your application by issuing the following command from the root folder of your application:

```
ojet serve
```

Packaging Web Application

Now that you have your web application ready and tested, next is package your application. From the root folder of your application, execute the command:

```
ojet build --release
```

Navigate to the "web" folder:

```
cd web
```

Execute the following command from the web folder to generate the TAR archive for the deployment:

```
tar -zcvf travvyWebviewApp.tgz *
```

Create Webview Service

Finally, you need to create a Webview service in your Digital Assistant environment to deploy the TAR archive created in the last step. Navigate to Components $f\equiv$. On this screen, you will see two tabs, namely, "Custom" and "Webview." Select the Webview tab and click the + Service button. Fill in the details as shown in Figure 8-10. Keep the "Service Hosted" toggle turned on and drag and drop the "travvyWebviewApp.tgz" from the web folder of your application to the "Package File" section.

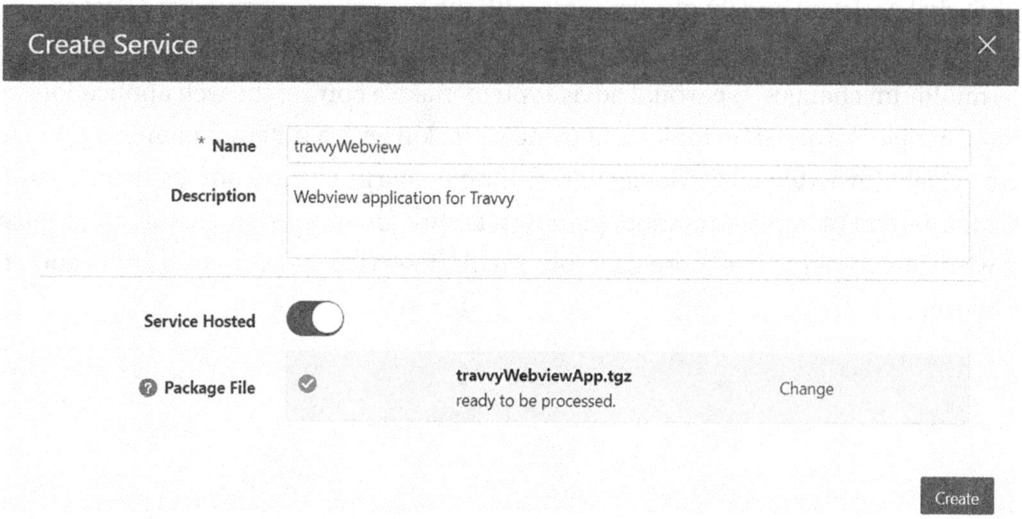

Create Service   ×

| * Name | travvyWebview |
| Description | Webview application for Travvy |

Service Hosted ⬤◯

❓ Package File ✓ **travvyWebviewApp.tgz** ready to be processed.   Change

Create

Figure 8-10. *Create Webview service*

Once done, click the Create button to create your Webview service. This process will take a few minutes, and once ready, you will receive a screen as shown in Figure 8-11.

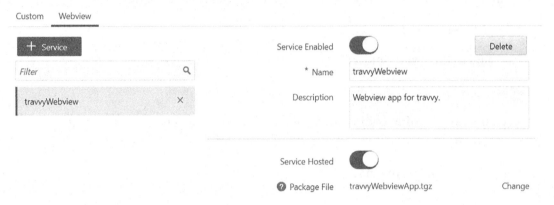

Figure 8-11. *Webview service*

Since you created this service with exact same name as mentioned in your System. Webview component's service property, no additional changes are needed. You can now test your end-to-end flow as described in Use Case section previously.

Webview Application Hosted Externally

In this section, we will explain to you how to host your web application outside of your Digital Assistant and invoke the same with the help of an intermediary service. For this part, you will be reusing the application created in previous sections with bare minimum changes. We would advise you to make a copy of the web application created earlier in a separate folder and clone your skill with a different name, say **"TravvyWebviewExternal."** Though in real-time scenario, you are not restricted to use a SPA for this kind of implementation, for now it is enough for the demonstration purpose. You will be using ngrok to expose your intermediary service and your web application to the Internet.

Create Intermediary Service

On your machine, create a folder with name "intermediaryService" to implement the service.

```
mkdir intermediaryService
```

Navigate to the folder using the following command:

```
cd intermediaryService
```

Configure this as a node project by issuing the command

```
npm init -y
```

This will generate a default package.json in your intermediaryService directory. You will be using Express to create this service. From the same command prompt/terminal, issue the following command:

```
npm install express --save
```

Create a file with name "index.js" in your intermediaryService directory and add the following code to it:

```
var express = require("express");
const bodyParser = require("body-parser");
var app = express();
app.use(bodyParser.json());

app.post("/travvyWebApp", (req, res) => {

    let requestBody      = req.body;

    let parameters = requestBody.parameters;

    let firstName = getWebviewParam(parameters, 'travellerFirstName', null);
    let callbackUrl = getWebviewParam(parameters, 'webview.onDone', null);

    //ngrok URL of the Web application
    let responseUrl = "https://d64ae6e3.ngrok.io?callbackUrl="+callbackUrl+
    "&firstName="+firstName;
```

```
    res.status(200).json({
        "webview.url": responseUrl
    })
});

function getWebviewParam (arrParams, key, defaultValue) {
    if (arrParams) {
        let param = arrParams.find(e => {
            return e.key === key;
        });
        return param ? param.value : defaultValue;
    }
    return defaultValue;
};

app.listen(4000, () => {
    console.log("Server running on port 4000");
});
```

In this service, you are reading input parameter (**travellerFirstName**) and callback URL sent by the skill with the help of **getWebviewParam** function. After that you are constructing a response URL which contains the URL of your web application, firstName (received as travellerFirstName from the skill), and callback URL from the Webview component received from your skill, for the web application. In the preceding code, you will need to replace web application URL "https://d64ae6e3.ngrok.io" which we will explain to you just in a while. You are passing this URL to your skill as a JSON object with in the "**webview.url**" parameter.

Web Application Implementation

Only change which you need to do in your previously copied application will be in main. js file. Navigate to the folder where you copied your web application. Locate your main.js file and replace the entire code with the following:

```
/**
 * @license
 * Copyright (c) 2014, 2019, Oracle and/or its affiliates.
```

```
 * The Universal Permissive License (UPL), Version 1.0
 */
'use strict';

/**
 * Example of Require.js boostrap javascript
 */

requirejs.config(
  {
    baseUrl: 'js',

    // Path mappings for the logical module names
    // Update the main-release-paths.json for release mode when updating
       the mappings
    paths:
    //injector:mainReleasePaths
    {
      'knockout': 'libs/knockout/knockout-3.5.0.debug',
      'jquery': 'libs/jquery/jquery-3.4.1',
      'jqueryui-amd': 'libs/jquery/jqueryui-amd-1.12.1',
      'promise': 'libs/es6-promise/es6-promise',
      'hammerjs': 'libs/hammer/hammer-2.0.8',
      'ojdnd': 'libs/dnd-polyfill/dnd-polyfill-1.0.0',
      'ojs': 'libs/oj/v7.0.1/debug',
      'ojL10n': 'libs/oj/v7.0.1/ojL10n',
      'ojtranslations': 'libs/oj/v7.0.1/resources',
      'text': 'libs/require/text',
      'signals': 'libs/js-signals/signals',
      'customElements': 'libs/webcomponents/custom-elements.min',
      'proj4': 'libs/proj4js/dist/proj4-src',
      'css': 'libs/require-css/css',
      'touchr': 'libs/touchr/touchr'
    }
    //endinjector
  }
);
```

```
/**
 * A top-level require call executed by the Application.
 * Although 'ojcore' and 'knockout' would be loaded in any case (they are
   specified as dependencies
 * by the modules themselves), we are listing them explicitly to get the
   references to the 'oj' and 'ko'
 * objects in the callback
 */
require(['ojs/ojcore', 'knockout', 'appController', 'jquery',
'ojs/ojknockout', 'ojs/ojformlayout', 'ojs/ojinputtext', 'ojs/ojlabel',
'ojs/ojbutton', 'ojs/ojradioset', 'ojs/ojvalidationgroup'],
  function (oj, ko, app, $) { // this callback gets executed when all
  required modules are loaded
    $(function () {

      const queryParameters = new URLSearchParams(window.location.search);

      /*
        Setting default values for the travellerFirstName if no value provided
      */
      self.firstName = ko.observable(queryParameters.get('firstName') !=
      null ? queryParameters.get('firstName') : ");
      self.lastName = ko.observable();
      self.address = ko.observable();
      self.zipCode = ko.observable();
      self.validId = ko.observable();
      self.idNumber = ko.observable();
      self.country = ko.observable();

      /*
        When the user submits or cancels the web form, control need to be
        passed back to the bot,
        For this a callback URL is passed from the webview to the web
        application. The parameter
        holding the information is "webview.onDone"
      */

      var webViewCallback = queryParameters.get('callbackUrl');
```

```
this.tracker = ko.observable();
self.buttonClick = function (event) {
  let userProfile = {};
  userProfile.firstName = self.firstName();
  userProfile.lastName = self.lastName();
  userProfile.address = self.address();
  userProfile.zipCode = self.zipCode();
  userProfile.validId = self.validId();
  userProfile.idNumber = self.idNumber();
  userProfile.country = self.country();

  var tracker = document.getElementById("tracker");

  if (tracker.valid === "valid") {
    // submit the form
    if (event.currentTarget.id === 'Submit') {
      userProfile.status = "success"
      //JQuery post call
      $.post(webViewCallback, JSON.stringify(userProfile));
    }
  }
  else if (event.currentTarget.id === 'Cancel') {
    let userProfile = {};
    userProfile.status = "cancel"
    //JQuery post call
    $.post(webViewCallback, JSON.stringify(userProfile));
  }
  else {
    // show messages on all the components
    tracker.showMessages();
    return;
  }

  const sleep = (milliseconds) => {
    return new Promise(resolve => setTimeout(resolve, milliseconds))
  }
```

```
        //Closes the browser tab the window is opened in. When closing the
          browser
        //tab, ensure the ajax call gets trough before the the tab closes.
          Thus adding
        //a "sleep" time
        sleep(500).then(() => {
          window.open(", '_self').close();
        })
        return true;
      }
      function init() {
        // Bind your ViewModel for the content of the whole page body.
        ko.applyBindings(app, document.getElementById('globalBody'));
      }
      // If running in a hybrid (e.g. Cordova) environment, we need to wait
        for the deviceready
      // event before executing any code that might interact with Cordova
        APIs or plugins.
      if ($(document.body).hasClass('oj-hybrid')) {
        document.addEventListener("deviceready", init);
      } else {
        init();
      }
    });
  }
);
```

Since your application is deployed outside of Digital Assistant, you will no longer have direct access to **getWebviewParam**. Well, for that purpose, you are using your intermediary service. Hence you are reading the parameters, which you received appended in the URL from skill. For this reason, you are now using **queryParameters** to get firstName and callback URL which are appended to web application invoked from your skill. The rest of your implementation remain intact.

Skill Implementation

Open your newly cloned skill TravvyWebviewExternal and replace its flow with that
mentioned in the following:

```
#metadata: information about the flow
#  platformVersion: the version of the bots platform that this flow was
   written to work with
metadata:
  platformVersion: "1.0"
main: true
name: TravvyWebviewExternal
#context: Define the variables which will used throughout the dialog flow
here.
context:
  variables:
    travellerFirstName: "string"
    appResponse: "string"
#states is where you can define the various states within your flow.

states:
  askTravelerInfo:
    component: "System.Text"
    properties:
      prompt: "To proceed with booking, please provide your first name."
      variable: "travellerFirstName"
    transitions:
      next: "webview"

  webview:
    component: "System.Webview"
    properties:
service: "travvyWebview"
      sourceVariableList: "travellerFirstName"
      variable: "appResponse"
      prompt: "Press 'Enter Traveler Information' to complete your booking"
      linkLabel: "Enter Traveler Information"
      cancelLabel: "Cancel"
```

```
    transitions:
      next: "evaluateWebviewResponse"
      actions:
        cancel: "onCancel"

  evaluateWebviewResponse:
    component: "System.Switch"
    properties:
      source: "${appResponse.value.status}"
      values:
      - "success"
      - "cancel"
    transitions:
      actions:
        success: "confirmBooking"
        cancel: "onCancel"
        NONE: "onCancel"

  confirmBooking:
    component: "System.Output"
    properties:
      text: "Booking has been confirmed for ${appResponse.value.firstName}
      ${appResponse.value.lastName}. Thanks for choosing Travvy!"
      keepTurn: false
    transitions:
      return: "done"

  onCancel:
    component: "System.Output"
    properties:
      text: "Sorry that you canceled your booking. Hope to see you back
      soon!"
    transitions:
      return: "done"
```

The entire implementation of your skill also remains the same except for the creation of service for the System.Webview component. In this case, you will create the "travvyWebview" service by toggling the switch for "Service Hosted" as shown in Figure 8-12.

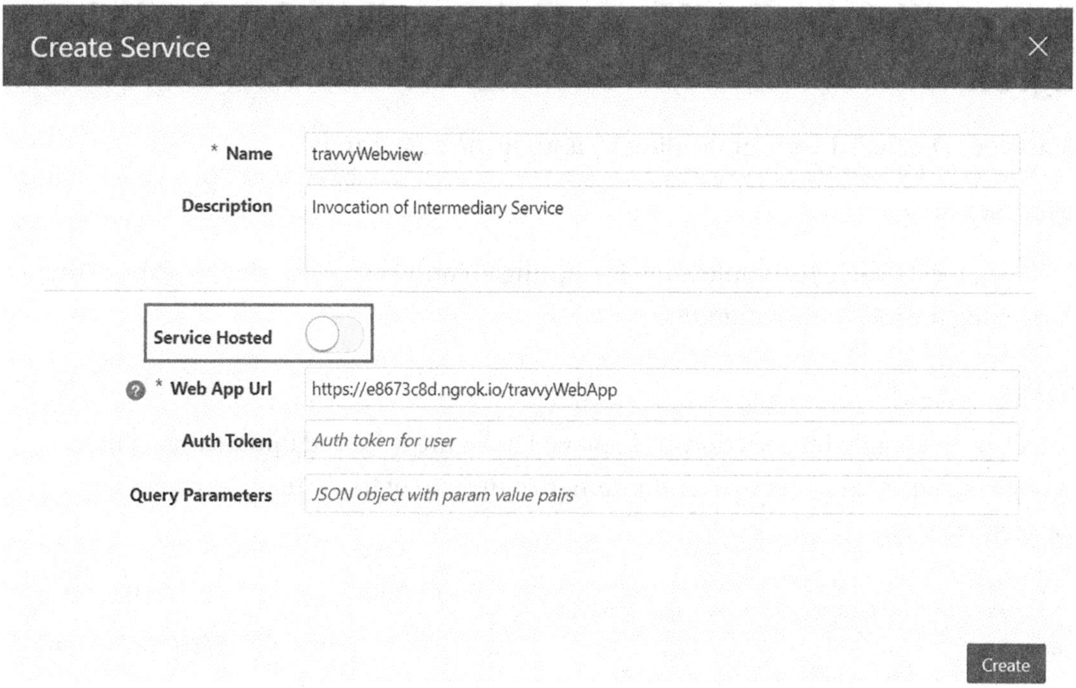

Figure 8-12.   *Create Webview service for externally hosted application*

You will need to update "Web App URL" after you generate the same in next section. Once completed successfully, service will look as displayed in Figure 8-13.

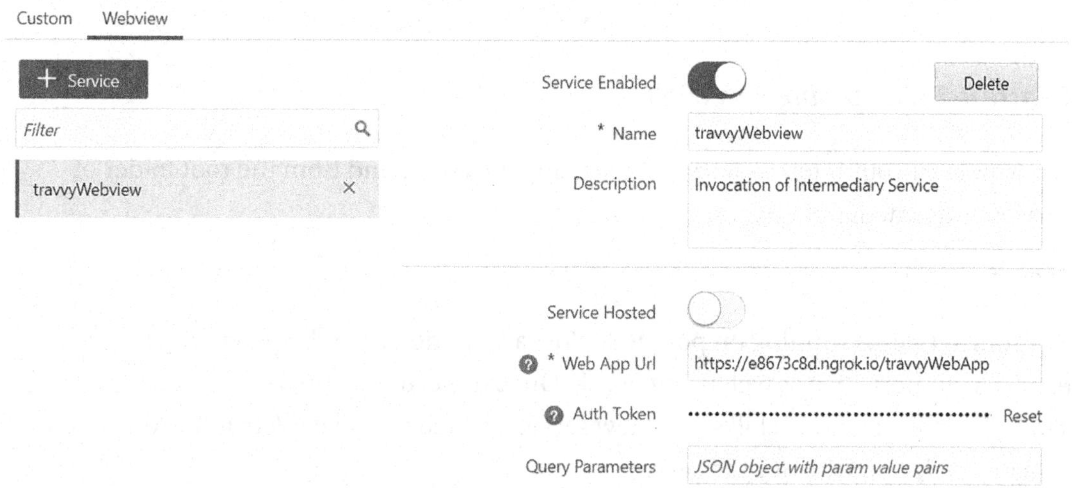

Figure 8-13.   *Service for externally hosted application*

Expose Web Application and Intermediary Service Using ngrok

Start your Oracle JET web application by issuing the command

```
ojet serve
```

Your local Oracle JET application is running on port 8000. You will first expose this port using the following command:

```
ngrok http 8000
```

Copy the highlighted section as displayed in Figure 8-14 and update this in index. js of your intermediaryService as mentioned at the end of the "Create Intermediary Service" section.

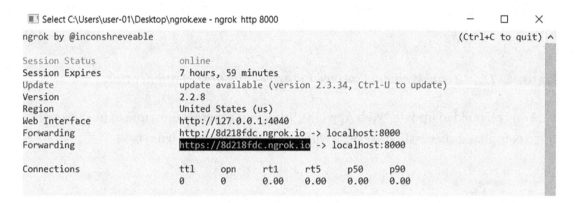

Figure 8-14. *Exposing port 8000*

Now start your intermediary service using the command from the root folder of service "intermediaryService":

```
node index.js
```

Your service is running on port 4000. Just as mentioned in the preceding text, expose your port 4000 as well using ngrok. Once done, update this URL in your TravvyWebviewExternal skill's Webview service, which is "travvyWebview". You need to update "Web App URL" for the service.

You can now test the end-to-end flow using your Skill Tester.

Summary

In this chapter, you were made familiar with Webview components, an alternative approach in conversation design to accumulate structured data from users. We explained you the Webview component architecture and two different approaches using which you can implement Webview component in your skill. Hope you find this topic interesting, and we will see you again in the next chapter.

CHAPTER 9

Building FAQs into Your Digital Assistant

Introduction

A very common use case for a Digital Assistant is the use of frequently asked questions (FAQs), also known as Q&A. Whenever the user types a phrase that matches a search term in the Q&A, the matching questions and answers are displayed to the user. A developed skill can act as the interface to these FAQs or other knowledge base documents. Frequently asked questions are common questions that are looking for an answer. "What are your opening times?" "Can I bring kids?" "Do you have vegan pizzas?" These FAQs often already exist on a company's web site and can easily be brought to your bot. The Q&A framework of Oracle Digital Assistant enables bots to answer general interest questions by returning one or more question and answer pairs. In this way, a bot can be used to surface FAQs or other knowledge base documents.

Our company, Extreme Hiking Holidays, has been running a web site for a long time. They have collected a whole bunch of frequently asked questions which are available from the web site. In this chapter, you will learn how to take these FAQs and their answers and add them to Travvy, the company's Digital Assistant, as Q&A.

How Does Q&A Work?

Before you learn how to implement Q&A, it is helpful to understand how Q&A works in Digital Assistant. There are some important configurations involved in setting up a good Q&A skill. How does intent resolution work, what is Q&A routing, and how do you build your Q&A flow?

© Luc Bors, Ardhendu Samajdwer, Mascha van Oosterhout 2020
L. Bors et al., *Oracle Digital Assistant*, https://doi.org/10.1007/978-1-4842-5422-6_9

Configuring Q&A Routing at Skill Level

For Q&A routing, Oracle Digital Assistant provides out of the box a set of commonly used words and opening phrases that indicate commands or questions in the Q&A Routing Config page. On top of these, any domain-specific content can be added. This configuration can be changed at the settings page of a skill, and it enables the router to discern between Q&A and intents. As you can see in Figure 9-1, there is a section for Transaction Config and a section for Question Config.

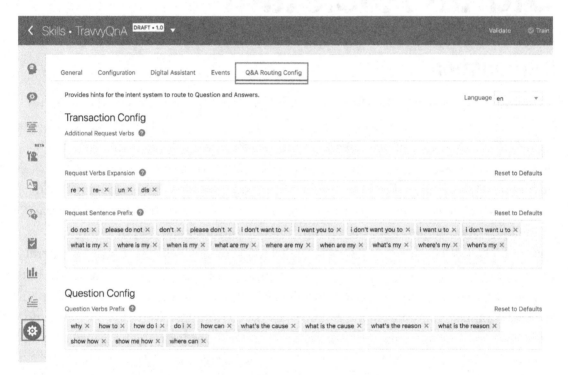

Figure 9-1. *Skill-level Q&A routing setting*

In the Transaction Config section, you can configure the skill to make it easier for the intent engine to recognize the utterance as a transaction (such as "**I want you to** book a trip," "**When is my** payment due"). These examples use the Request Sentence Prefix. Also, by default, there is a wide set of Request Verbs, such as "order," "cancel," "check," and more. You can add Request Verbs that are specific for a use case. As you can see from Figure 9-1, the built-in set already contains the most commonly used verbs and prefixes, but you can always add your own to cater to specific transactions such as "book" or "check in" which would be typical for our travel business.

In the Question Config, you can create a set of words that indicate that a user is asking a question ("**how do I** book a trip," "**how can** I cancel my vacation") and not requesting a transaction. These can also be extended with specific Question Verbs.

How Q&A Works at Runtime

At runtime, the bot's behavior is influenced by the configurations as described in the previous section to check. On top of that, there are two important properties involved in Q&A routing:

- confidenceTreshold: The skill uses this property to steer the conversation by the confidence level of the resolved intent. This property is set at the Skill-level configuration.

- Q&AMinimumMatch: Sets the minimum and maximum percentage of tokens that any Q&A pair must contain in order to be considered a match. The value of this property can be set at the System.Intent components in the flow editor.

Tip   The Q&AMinimumMatch setting defaults to 50%/25%. While it does not make sense to set it to, for example, 90%, you may want to tune the QnA findings by, for example, setting it to only 50% or only 30%. The 50%/25% default means that if the 50% is not matched, the search falls back to 25%, so going for a weaker condition.

The combination of these two properties leads to the final resolution and the outcome of the flow. When the user inputs a phrase, there are a couple of possible scenarios. Let's have a look at some examples:

- Over the confidenceTreshold and does not meet Q&AMinimumMatch: In this case, the intent is resolved, and the user is shown the bot's reply to this intent (Figure 9-2).

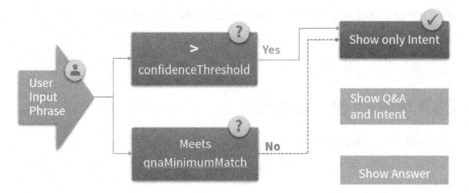

Figure 9-2. *Situation 1*

- Below the confidenceTreshold and does meet Q&AMinimimMatch: In this case, the bot will reply to the user the answer that belongs to the asked question (Figure 9-3).

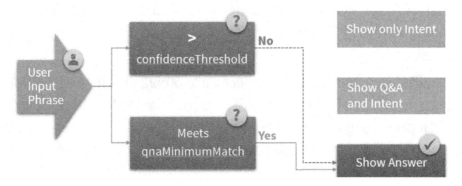

Figure 9-3. *Situation 2*

- Below the confidenceTreshold and does not meet Q&AMinimimMatch: The bot will not know how to resolve the intent and will show the user the unresolved intent prompt (Figure 9-4).

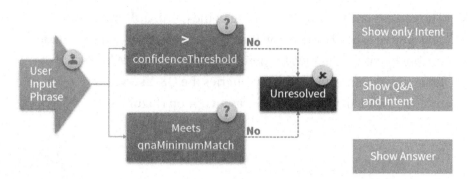

Figure 9-4. *Situation 3*

- Over the confidenceTreshold and meets Q&AMinimimMatch: In this
  scenario, there are a couple of possible outcomes:

  - The first one depends on the use of imperative verbs like "buy,"
    "check," "order," "confirm," and "cancel." These words indicate a
    transaction, not a question, so whenever the utterance uses an
    imperative, the bot will reply with the intent only. If the utterance
    does not contain an imperative, the bot will show both the Q&A
    and the intent (Figure 9-5).

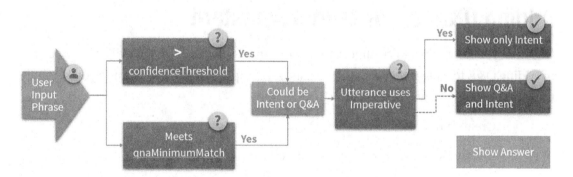

Figure 9-5. *Situation 5*

- The second one depends on setting of the QnASkipIfIntentFound. This property can be set at the System.Intent level. When set to **true**, the bot bypasses Q&A when there's an intent match. When set to **false** (the default value), the bot queries the Q&A server with user utterance and also presents Q&A as an option (Figure 9-6).

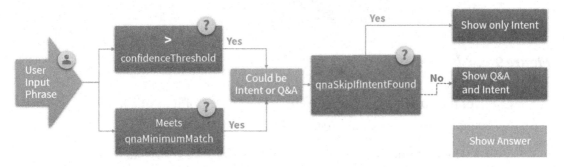

Figure 9-6. *Situation 5*

In the remainder of this chapter, you will learn how to use these properties in a Q&A skill for the Travvy Digital Assistant.

Adding Q&A to Our Digital Assistant

Now that you know how to configure Q&A, it is time to develop a Q&A skill. The process of adding Q&A to a Digital Assistant consists of multiple steps. We will guide you through these steps, and they will be explained individually:

- Create a new skill to hold the Q&A.

- Enable the Q&A capability for the skill.

- Load the Q&A source file, and optionally edit, add, or delete more questions.

- Train the bot with the Q&A trainer.

- Test the new Q&A capability by entering questions into the Q&A Tester.

- To use Q&A in a bot dialog flow, we need to configure the System. QnA.

- Add a state for the *none* action, and set the transition back to the Q&A component when the question is answered by adding Q&A for the next transition.

To get started with Q&A, a new skill has to be created (Figure 9-7).

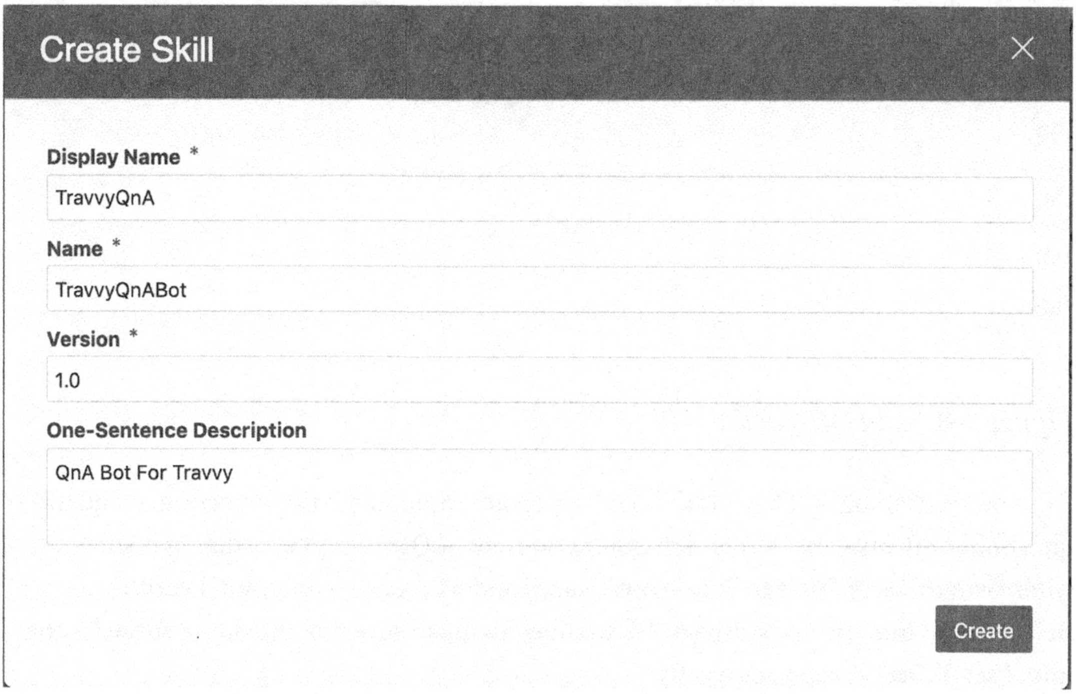

Figure 9-7. *Creating a new Q&A Skill*

After you click Create, the skill editor is opened, and everything is ready to add the questions and answers, write the flow, and test the Q&A Skill. After initial creation of the skill, the bot builder does not know that this particular skill should be created as a Q&A skill. In order for the skill to become a Q&A skill, Q&A needs to be enabled by clicking the Q&A icon  in the navigation bar. Then click **Add Q&A** (Figure 9-8).

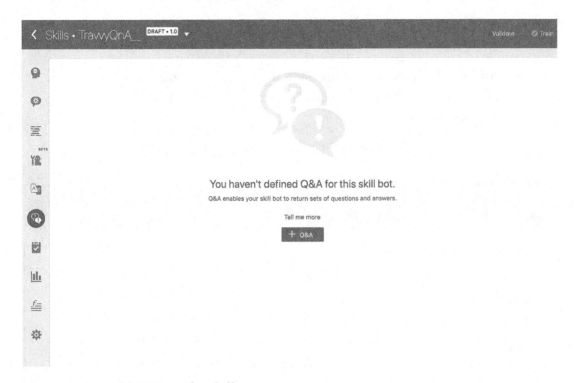

Figure 9-8. *Add Q&A to the skill*

Now everything is set up, and the next step is to upload the csv that contains all the questions and answers for this skill. After you click "+ Q&A Source," a dialog is shown where you can set language, locale, and name and where you can upload your csv. Before we go into the uploading itself, we need to make sure that the data source for the initial set of Q&A is set up correctly.

The Q&A Data File

The Q&A data file is a file that contains all the question and answer pairs that you initially need for your skill. Typically, this file would bundle your FAQ in a format that can be handled by Oracle Digital Assistant. The file has to meet some criteria in order to be uploaded to the Digital Assistant Cloud:

- This file has to be a UTF-8 encoded CSV file.

- The file must have header row with category_path, questions, and content. Below that row the real questions and answers can be added. An example can be seen in Figure 9-9.

A	B	C	D
category_path	questions	content	
Trip destinations	Where can I find travel destinations? Do you have a list of destinations? Can I get an overview of all destinations? What are possible destinations for my trip?	You can find all our travel destinations on the website	
Trip destinations	Do you also offer hiking trips in France? Can I book a trip to another country than Canada? I'm looking for a hike in Austria What about a hiking trip in Nepal?	In our travel program we offer hiking trips in Canada, the USA and in Switserland	
	What to expect from a hike in Switzerland? Can you give me some details about hiking in Canada?		

Figure 9-9. *Example Q&A data source file*

The category_path contains the category (or categories) for a given question and answer (Q&A) pair. Such a Q&A pair can belong to more than one category. If that is the case, each category should be entered on new line. If there is a hierarchy in the categories, a forward slash should be used:

- travel

- travel/insurance

- travel/itinerary

In the questions column, you enter the questions that are used. Each versions of a question should be entered on a new line. The first question, known as the canonical question, is the question that displays in the bot's message by default. The subsequent questions in the column are alternative versions.

You could add all questions and answers manually in the cloud UI, but using a spreadsheet for entering data is a much more convenient way of doing this. Using the spreadsheet for you initial bulk load of Q&A will save you time. If at a later stage there is a need to add extra individual questions and/or answers, these can always be added manually in the UI after uploading the csv.

Tip   The more questions you add for each answer, the more likely it is that relevant Q&A pairs are returned to the user of the skill. Add between 2 and 5 question for each answer. This will improve accuracy.

When adding the questions and answers, you have to make sure enough questions are provided to the skill in order to recognize all different ways that a user can ask the same question. The best way to do this is to vary the verb and interrogatives like "how," "what," and "where," but be very consistent with the answer's topic.

A question like "Where can I find travel destinations?" could be asked in different ways such as:

- Do you have a list of destinations?

- Can I get an overview of all destinations?

- What are possible destinations for my trip?

It is obvious that these questions should always return the same answer. When you are satisfied with the content of the Q&A csv, you can now add it to your skill. This upload process can be started by invoking the "+ Q&A Source" button as shown in Figure 9-10.

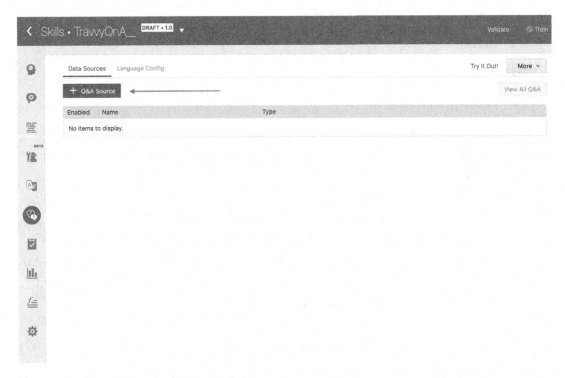

Figure 9-10. *Add new Q&A source*

An upload dialog will appear (Figure 9-11) where you can pick the language and locale for the Q&A source and give it a name. Next the csv file can be dragged onto the dialog in order to upload it. Once all fields are entered, click Create, and the csv will be uploaded and added to your Q&A.

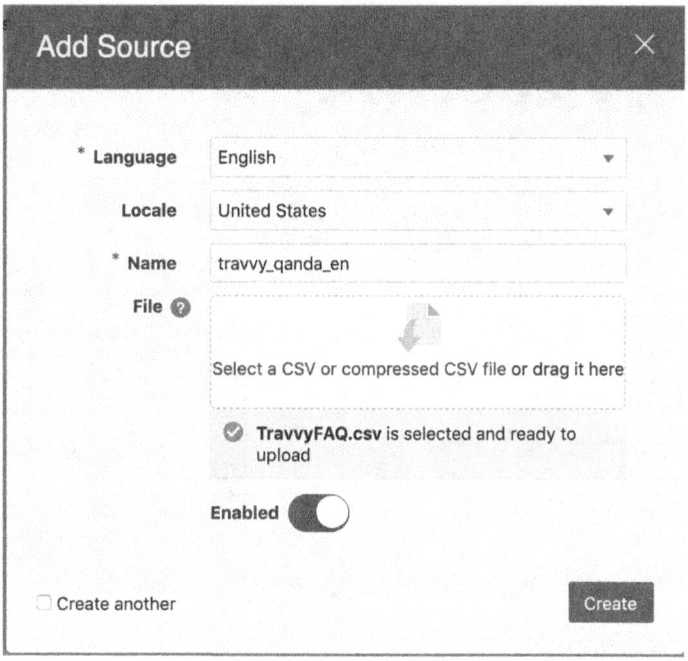

Figure 9-11. *Adding a source for Q&A*

Tip   If, after uploading, you decide to edit the data source file and replace the content, be aware that it will replace all online content, with the content from the new csv file. Any changes you did in the cloud will be overwritten. It is better to do changes in the csv instead of in the cloud. If you make changes in the cloud, make sure that, before uploading a new file, you export the cloud Q&A content.

All the questions and answers in the csv are now uploaded to the Skill Editor. In the list, you can see the file that you just uploaded.

One more thing you can do to improve the quality of the Q&A is to add language-specific configuration. Here you can add ignored words, abbreviations, and synonyms. For instance, for "trip," there are synonyms such as "journey," "holiday," "voyage," and more.

By simply adding them here as synonyms, they can all be used to resolve to the same question (Figure 9-12).

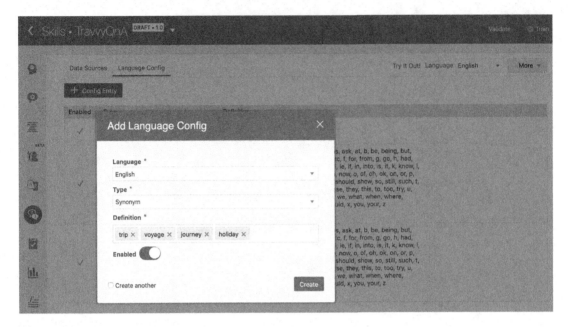

Figure 9-12. Add synonyms

Note   In this chapter, we will not go into the details of multilanguage. The specifics of multilanguage were discussed in Chapter 7.

When you click "View All Q&A," you will see a list that shows the exact content of that file (Figure 9-13).

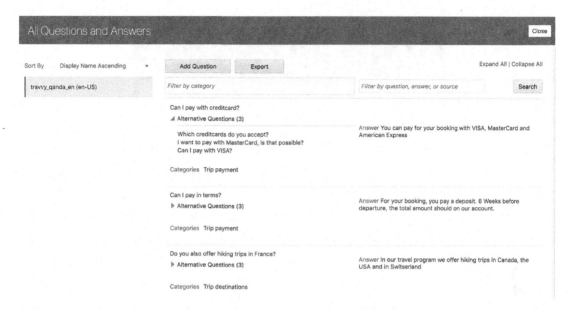

Figure 9-13. *All Q&A for Travvy*

This is also the place to edit the Q&A and to make an export if needed. It is now time to train the Q&A.

Testing the Q&A

In order to make sure that the Q&A works as expected, it has to be tested, but before we test the Q&A, we must train the Q&A (Figure 9-14). This is a simple task that can be executed in the training section of the bot builder:

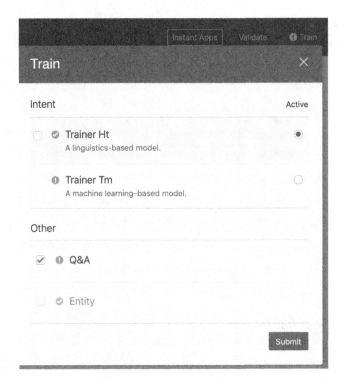

Figure 9-14. *Training the Q&A for the skill*

Once this is done, the tester can be invoked by clicking "Try It Out" to start testing the Q&A for Travvy. This can be done on a "one by one" basis or in batch. For Travvy, we will stick with the "one by one." Let's check what happens when we enter "How do I pay my trip?" (Figure 9-15).

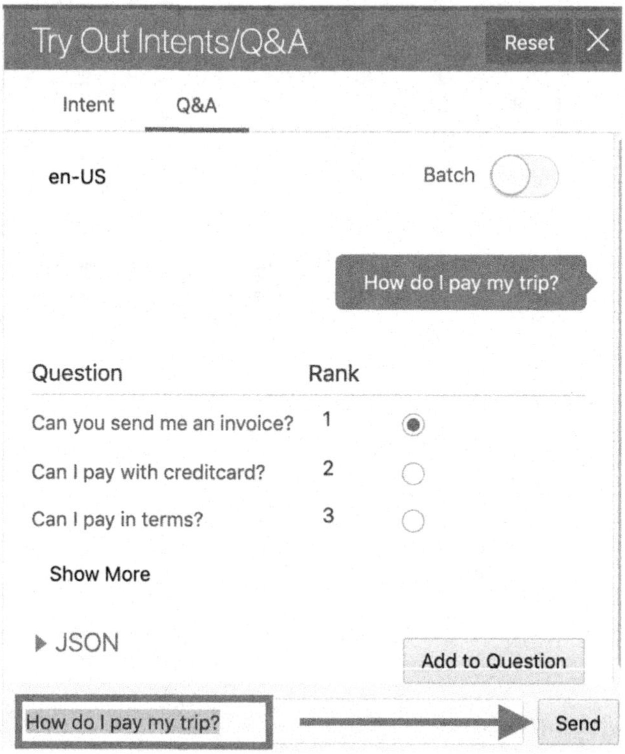

Figure 9-15. *Testing the Q&A*

The Q&A capability handles the responses to the utterances that were entered. It consists of a set of questions which represent the best fit. In our test case, there are three possible questions that relate to the user's utterance, all belonging to the "trip payment" category. Outside the tester, in the real "chatbot," these answers will be shown in a kind of carousel, so the user can browse all matching questions in order to find the one he is looking for.

Now let's have a look at what happens if we change that to "How do I pay my booking?" (Figure 9-16).

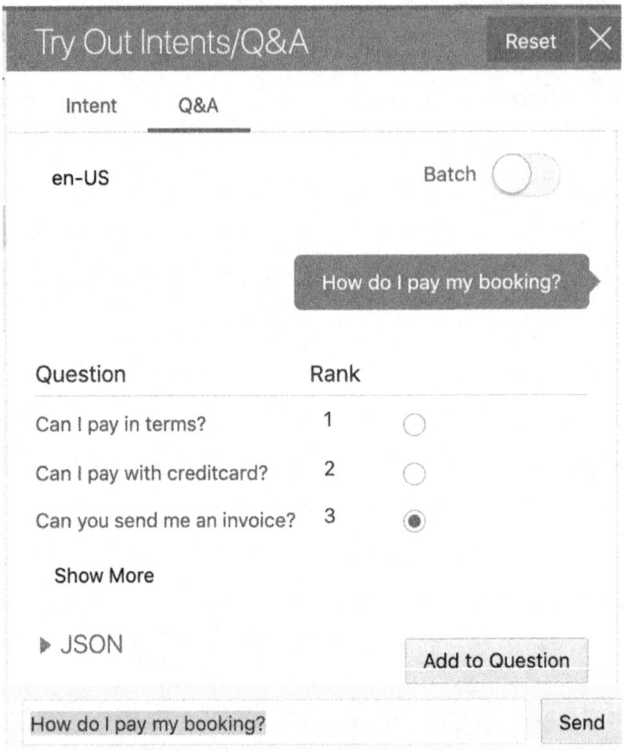

Figure 9-16. *Testing the Q&A*

The Q&A capability now picks a different question and a different order. Obviously all questions somehow relate to the entered utterance. If you want to increase the weighting of a question and answer for an utterance, you can use the "**Add to Question**" button to add your utterance to a specific Q&A question.

Adding Q&A to the Dialog Flow

The final step in setting up this Q&A skill is to enter the Q&A flow in the flow editor. If you open the flow editor, you will see that it already contains code. Everything below the "states" keyword can be removed, because it is not needed for this Q&A bot. Also you can remove the comments if you like to. With a clean editor, it is now easy to set up the Q&A flow.

This "Q&A only" skill will have two states:

- Q&A: The state that allows for the questions to be entered and the answers to be shown.

- Unresolved: As always, a skill needs an "unresolved" state to make sure the conversation does not crash when the Digital Assistant does not understand what the user wants.

The "Q&A" state here is the main state for the Q&A, and the next state should always be "Q&A," thus routing back to Q&A when an answer is returned. Therefore, the configuration is as follows:

- component: "System.Q&A"

- transitions:

    - actions:

        - none: "unresolved"

    - next: "travvyQnA"

The final result is shown in the following code:

```
metadata:
  platformVersion: "1.0"
main: true
name: "TravvyQnA"
context:
  variables:
    greeting: "string"
    name: "string"
    terminateChoice: "string"
states:
  travvyQnA:
    component: "System.QnA"
    transitions:
      actions:
        none: "unresolved"
      next: " travvyQnA"
```

```
unresolved:
  component: "System.Output"
  properties:
    text: "Sorry, I did not find any match. Can you rephrase the
    question?"
  transitions:
    return: "done"
```

Let's see what happens when we ask "how can I pay my voyage" in the Skill Tester.

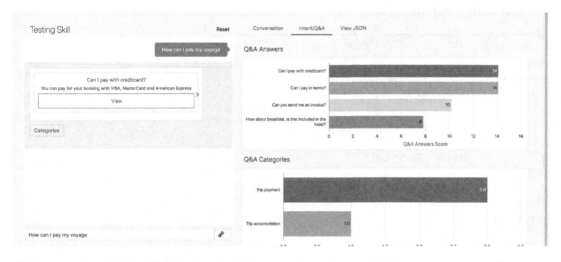

Figure 9-17. *Default behavior in tester*

You see at the right side of Figure 9-17 that there are four Q&A sets selected by the Q&A. All four are also displayed in the carousel where you can use the arrow buttons to browse the options. This behavior can be changed by tweaking the properties of the system.QnA component. To reduce the number of Q&A pairs presented to the user, you can use the matchListLimit property.

In the following listing, this property is set to "1":

```
travvyQnA:
  component: "System.QnA"
  properties:
    matchListLimit: 1
```

```
transitions:
  actions:
    none: "unresolved"
  next: " travvyQnA"
```

The behavior will change, and you can see this in the tester (Figure 9-18):

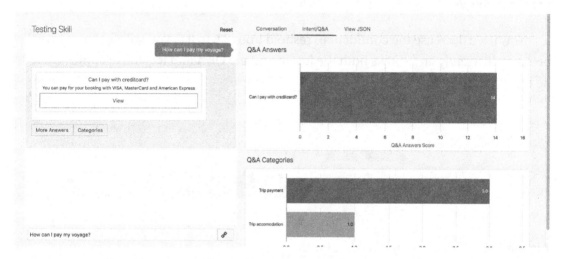

Figure 9-18. *Behavior with matchListLimit=1*

Only the first option is displayed to the user, and he gets an extra option to display "More Answers."

There are more options that you can use to influence the behavior of the Q&A such as the aforementioned QnASkipIfIntentFound and the confidenceTreshold.

To show what happens when you mix Q&A with Intents and play with the available options, we added a simple intent (PayTrip) to the Q&A skill.

Note   For the Travvy Digital Assistant, we will keep the Q&A in a clean separate skill without intents. At a later stage, we will add this skill to the Digital Assistant.

First we will have a look at what happens with the confidenceTreshold. We will use the utterance **"I need to fulfill the invoice,"** which can relate to Q&A or to the PayTrip intent. When the confidenceTreshold is set to 0.6, the skill will immediately show the Q&A that it finds (Figure 9-19), because the Intent score percentage is 0.49, which is less than the confidenceTreshold.

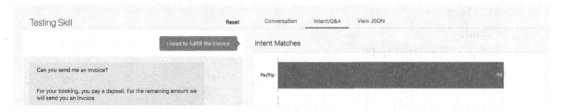

Figure 9-19. *Less than confidenceTreshold*

Now if we decrease the confidenceTreshold from 0.6 to 0.5, you will see that the skill behaves differently and shows both the intent and the Q&A and lets the user pick what he wants (Figure 9-20).

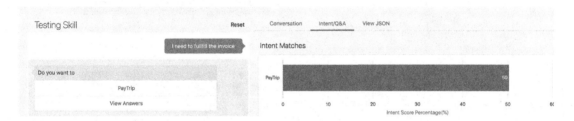

Figure 9-20. *Over confidenceTreshold*

Now finally we can change the QnASkipIfIntentFound from false to true and see what the effect is (Figure 9-21).

```
Intent:
  component: "System.Intent"
  properties:
    variable: "iResult"
    qnaEnable: true
        qnaSkipIfIntentFound: true
  transitions:
    actions:
      unresolvedIntent: "unresolved"
      payTrip: "payTrip"
      qna: "travvyQna"
```

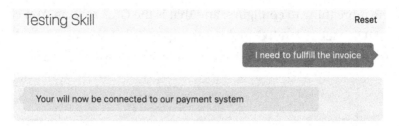

Figure 9-21.   qnaSkipIfIntentFound is set to true

Now let's set QnASkipIfIntentFound back to false. When the user says "**I need to pay my voyage**," the skill shows both options because there is an intent scoring above the confidence threshold and a matching question, but ranks the intent first (Figure 9-22).

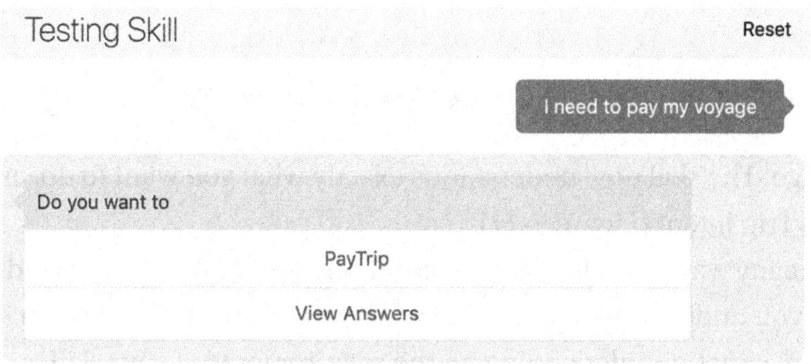

Testing Skill   Reset

I need to pay my voyage

Do you want to

PayTrip

View Answers

Figure 9-22.   Intent is showed first

When the user uses one of the "Question Verbs or prefixes" as defined in the Q&A Routing Config, this indicates that a user is asking a question. The link for the Q&A response is shown first as it gets "priority" over the link for the intent response (Figure 9-23).

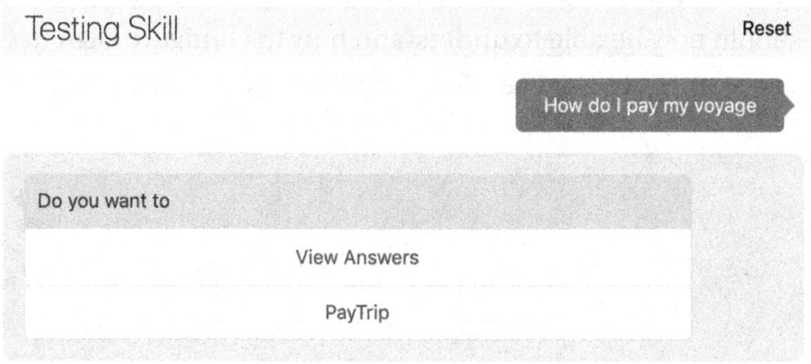

Testing Skill   Reset

How do I pay my voyage

Do you want to

View Answers

PayTrip

Figure 9-23.   Q&A shown first

There is one more thing to configure, and that is the Q&A routing. You can add custom "**Request Sentence Prefix**" to the routing section. Let's see what will happen if you add "**I want to**" to that section and ask the skill "I want to pay my trip."

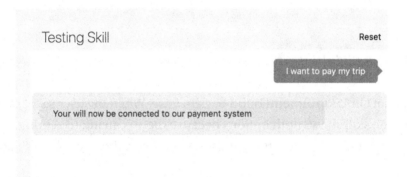

Figure 9-24. *Intent shown after configuring Q&A routing*

There you go. The skill now understands exactly what you want to do, and it will resolve the PayTrip intent (Figure 9-24).

There are many ways to influence and configure the Q&A behavior, and besides it is really key that you understand all the options, as you will have to use them in different ways in different situations, depending on the requirements of your skill.

Summary

Frequently asked questions are common questions that are looking for an answer. They often already exist on a company's web site and can easily be brought to your bot. In this chapter, you learned how to develop a Q&A skill and how to use it in your Digital Assistant. You should now be able to understand how to configure Q&A in Oracle Digital Assistant and how to improve Q&A sets for an optimal user experience.

PART V

Advanced Topics

CHAPTER 10

Extending Your Digital Assistant with Custom Components

Introduction

Welcome to the chapter of Custom Components. This chapter will focus explicitly on Custom Components for Oracle Digital Assistant (ODA). During the course of Chapter 5, you were introduced to Custom Components briefly. Presuming that you have an overview about custom components, this chapter will explain Custom Components in detail. Custom components are those components which you can handcraft by yourself, which are specific to your need, and most importantly whose behavior you have a complete control over. You will start with the basics of creating custom components in your local environment, debugging them locally, hosting them, and, finally, using them in your skill's dialog flow.

As it was mentioned earlier, any task or action that you want to perform in a Digital Assistant or, to be more precise, in a Skill, you will need a component for that. Consider a very simple use case where you ask the user to enter his/her name and then you display a customized greeting message specific to user name. You will be using at least two components to achieve this. First, you will use a component to display a prompt that asks the user to provide his/her name. And then using a second component, you will display a customized greeting message with user name based on the input provided.

Every action and bot response, be it a prompt for an input or a message to display, requires a state in the dialog flow. Every state will invoke one and only one component to perform the task. Though Oracle Digital Assistant provides you a variety of built-in components out of the box which you can use to build your bot flow, you will often come across situations where built-in components are just not enough to perform specific tasks.

© Luc Bors, Ardhendu Samajdwer, Mascha van Oosterhout 2020
L. Bors et al., *Oracle Digital Assistant*, https://doi.org/10.1007/978-1-4842-5422-6_10

Specific tasks, like backend integrations, where you need to invoke any REST service to get desired information from backend systems and then perform data manipulation before showing the result to user, will need custom components. This cannot be done using built-in components only. This is a typical use case for custom components where you define your custom logic inside a custom component using JavaScript, host the component, and then invoke it from your dialog flow to get the desired result. While you work with Travvy, you might want to retrieve user-specific information, like full name, age, address, and so on using a REST service.

Another scenario, which you can consider for using a custom component, is when you want to implement some complex logic for your skill, but you realize that the job cannot be accomplished using Apache FreeMarker. In the context of this book, consider a use case for Travvy where you only want to allow users to book a trip if they are grown up, that is, if their age is greater than 18. If user's age doesn't match your criteria, then you display a message stating that he/she is not eligible to book a trip using Travvy. Such type of complex logic is not very convenient to implement directly in bot flow but can be done effectively using custom components. Having said that, you can outline the following benefits of using custom components based on what we have discussed so far.

Benefits of Using Custom Components

- Custom components keep your dialog flows short for use cases that otherwise require multiple states to implement.

- Custom components, if deployed as an API using Oracle Mobile Hub (refer to Figure 10-13 under section "Access Custom Component from ODA Instance" where you can select the Oracle Mobile Cloud option), can be used across different skills, which can be even spread across various ODA instances.

- As custom components are developed using Node.js, you can leverage public and free third-party Node libraries.

Use Cases for Custom Components

- By default, Oracle allows you to use Google and Microsoft translation services to provide multilingual support for your bot. But if you wish to use any third-party translation services for your bot, you can also do that using custom components.

- You may also think about using custom components if you wish to implement security for your bot or in any specific section like for API invocations.

And the list doesn't end here. As you start working with custom components, you will yourself come up with many more benefits and use cases of using them in your specific scenarios.

Now that you hold a basic understanding about the custom components and where you should consider using custom components, it's the time to start working with them. In next few steps, you will see how easy it is to create your first custom component and to start implementing your custom logic. We would also suggest you acquire some knowledge of JavaScript language in case you don't happen to have it already. That will come a lot handy going forward, and you will notice the difference while working on custom components. But first, you will have to set up your local environment for development. This would allow you to develop your custom components on your own machine, debug, and test before you actually deploy them on the cloud.

Environment Setup for Custom Component

There are a few prerequisites you need to install on your local machine in order to start working with your custom components. First and foremost, you need to install node on your local machine if you don't already have it installed. You can download Node from the following link:

```
https://nodejs.org/en/download/
```

Next thing in the list is to download Oracle Bots Node.js SDK. Once you have installed Node, you can simply install Oracle Bots Node SDK globally using the following command:

Windows

```
npm install -g @oracle/bots-node-sdk
```

Mac

```
sudo npm install -g @oracle/bots-node-sdk
```

Next, you need to download a tunneling software. We are going to use ngrok. This will make your life a lot easier while debugging and testing your custom component by allowing you to access the custom component located on your development computer directly from ODA. You can download ngrok from the following link:

```
https://ngrok.com/download
```

Using ngrok, you can expose a specific port from your machine over the Internet. This is discussed in detail in section "Expose Custom Component over the Internet."

Finally, download a code editor for JavaScript development based on your preference. You may consider using Microsoft Visual Studio Code.

Getting Started with Custom Component Development

Assuming that you followed steps mentioned in the previous section, your local environment should be ready for custom component development. As discussed in the beginning of this chapter, in next few steps, you will create a custom component to validate the user's age.

Navigate to the directory where you want to create your custom component and open a command prompt from there. Ensure that the directory where you want to create your custom component and any directory you create inside your custom component should not contain any blank space. See Figure 10-1 for reference.

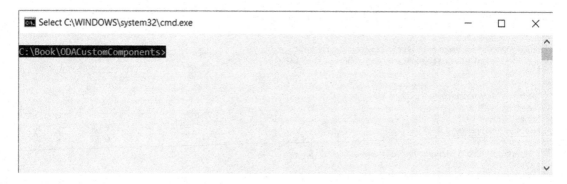

Figure 10-1.   *Custom Component location*

From the preceding location, you will create a directory with name ageValidatorCC. Use the following command in the command line:

```
mkdir ageValidatorCC
```

Once created, navigate to the directory ageValidatorCC by using the following command:

```
cd ageValidatorCC
```

The next step is to configure your project directory, that is, ageValidatorCC directory, as a node project. Use the following command:

```
npm init -y
```

This will generate a default package.json inside the directory ageValidatorCC. You can later update this package.json file with your project-specific details. Refer to Figure 10-2.

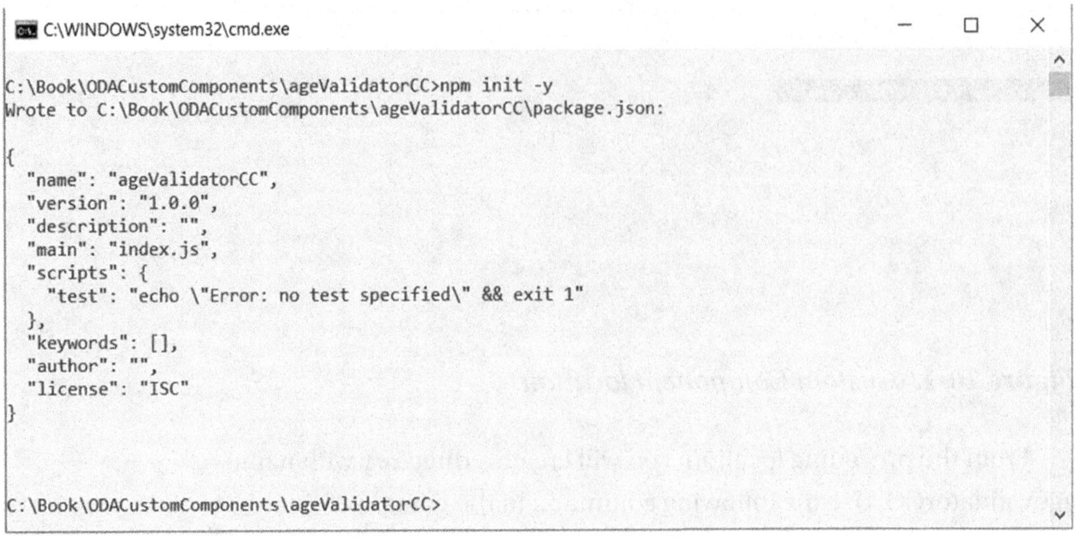

```
C:\WINDOWS\system32\cmd.exe                                                —    □    ×

C:\Book\ODACustomComponents\ageValidatorCC>npm init -y
Wrote to C:\Book\ODACustomComponents\ageValidatorCC\package.json:

{
  "name": "ageValidatorCC",
  "version": "1.0.0",
  "description": "",
  "main": "index.js",
  "scripts": {
    "test": "echo \"Error: no test specified\" && exit 1"
  },
  "keywords": [],
  "author": "",
  "license": "ISC"
}


C:\Book\ODACustomComponents\ageValidatorCC>
```

Figure 10-2. *Configure a node project*

Next, add Oracle Bots Node SDK to your project directory using the following command:

```
npm install --save-dev @oracle/bots-node-sdk
```

If you navigate to directory ageValidatorCC now, you will notice that the directory contains package.json which was created in the last step and also newly added node modules inside node_modules directory.

Next you need to add a JavaScript file where you can write your custom logic. For this purpose, use the following command:

```
bots-node-sdk init -c AgeValidator
```

Once done, you will see a message stating "Custom Component package 'ageValidatorCC' created successfully!" This will create component package inside your directory and add dependencies. Refer to Figure 10-3.

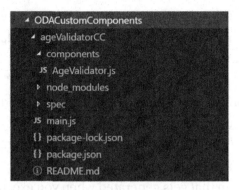

Figure 10-3. *Custom Component package*

Congratulations! You have now successfully created your first custom component. If you have prior experience of working with node projects, then you must be familiar with files displayed in Figure 10-3. But if not, we will briefly go through the files which are mainly of your interest:

package.json: This file contains your project-specific metadata information. This includes reference to the entry point of the project, which in this case is main.js, and JavaScript dependencies for your project. To learn more, refer to the following link:

```
https://docs.npmjs.com/files/package.json
```

main.js: As mentioned before, this file is the default entry point for your custom component project. You will notice this file refers to the component's directory where you will be adding your custom component implementation. Refer to the following code of main.js:

```
module.exports = {
  components: [
    './components'
  ]
};
```

Even if you implement more than one component inside your component directory, you can simply access specific component by specifying the component name. You don't have to modify anything in your main.js to access that. For example, like you have added AgeValidator component, you can also add another component GetUserProfile to retrieve user profile information inside the component's directory.

Another point which becomes important in this context is to create some sort of logical group at the time of component creation. For instance, you might want to put AgeVaildator and GetUserProfile profile components in different directories inside

247

components to isolate to different implementations. One thing which you might want to change is to give a generic name to your parent folder like odaCustomComponentCC which in this case was ageValidatorCC.

Though this is not a must, it is a good practice to adhere. After all it will help you as well as your fellow team members to understand the code implementation and pinpoint to a specific component in case of any issue.

AgeValidator.js: This is the JavaScript implementation of your custom component which resides under components directory. Figure 10-4 shows you the default custom component implementation that gets generated when you execute "*bots-node-sdk init -c AgeValidator*".

```js
JS AgeValidator.js ✕
 1   'use strict';
 2
 3   module.exports = {
 4     metadata: () => ({
 5       name: 'AgeValidator',
 6       properties: {
 7         human: { required: true, type: 'string' },
 8       },
 9       supportedActions: ['weekday', 'weekend']
10     }),
11     invoke: (conversation, done) => {
12       // perform conversation tasks.
13       const { human } = conversation.properties();
14       // determine date
15       const now = new Date();
16       const dayOfWeek = now.toLocaleDateString('en-US', { weekday: 'long' });
17       const isWeekend = [0, 6].indexOf(now.getDay()) > -1;
18       // reply
19       conversation
20         .reply(`Greetings ${human}`)
21         .reply(`Today is ${now.toLocaleDateString()}, a ${dayOfWeek}`)
22         .transition(isWeekend ? 'weekend' : 'weekday');
23
24       done();
25     }
26   };
27
```

Figure 10-4. *Custom component default implementation*

As you can see, AgeValidator, by default, exposes two different functions. They are metadata and invoke. Let's discuss them in a bit more detail here:

metadata: (): As you can see, this function holds the metadata information of your component. metadata function constitutes of three different attributes.

- name: This is the unique name to identify your component.

- properties: This is an object which holds the parameters which are essential for your component execution.

- supportedActions: These are actions supported by your component, which you can conditionally dispatch from your component. Based on these actions, you can define state transition actions in your bot flow.

invoke: (conversation, done): invoke function accepts two argument, namely, conversation and done. conversation is a reference to custom component SDK. You can use its functions to read the messages which come out from the bot to your custom component and write component responses based on those messages. For example, assume that you defined a variable as userAge along with its type in properties section, which you are expecting as an input parameter for you custom component. Once you pass userAge value from your skill, to get that value in your custom component, you can use the following code:

```
var givenAge = conversation.properties().userAge;
```

Likewise, you can also define a variable (say ccResult) in properties section along with its type. Then use this variable to pass a value to your bot by setting it as follows:

```
conversation.variable(ccResult, ccCalculatedVal);
```

You will invoke done(), which is a callback, as soon as your custom component finishes its processing and you want to pass the handle to the bot dialog flow. For example, you can invoke done() right after setting the calculated value for the variable as explained in the preceding code. Thus, it would look like as follows:

```
conversation.variable(ccResult, ccCalculatedVal);
done();
```

Run Your Custom Component Locally

Now that you have the base implementation of your custom component ready, the next step is to run it locally. Refer to Figure 10-3 to see the directory structure of the project.

Navigate to the parent directory ODACustomComponents which contains the custom component ageValidatorCC, to start a node server. Refer to Figure 10-4.

Figure 10-5. *Custom Component parent directory*

Using a command prompt/terminal, execute the following command as displayed in Figure 10-6 to start the node server:

```
bots-node-sdk service ageValidatorCC
```

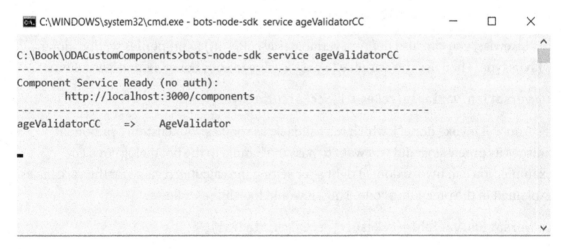

Figure 10-6. *Start node server*

Now that your component is running on your local machine, you can check it by navigating to URL `http://localhost:3000/components` on your browser. It will look as shown in Figure 10-7.

Figure 10-7. *Custom Component running on local machine*

In the preceding text, you can see various metadata information from your component based on initial implementation. As you will change this implementation for AgeValidator in later part of this book, you will notice that these metadata information will also change according to your new implementation.

Expose Custom Component over the Internet

So far, you have successfully created your first custom component on your local machine and also started it on a local node container using Oracle bots-node-sdk. You also tested your running component using a web browser. You are one step away from accessing your custom component from an ODA instance. For that you will need to expose your IP over the Internet, so that you can access your service using that IP. You will now use ngrok to accomplish that.

Navigate to the directory where you have downloaded and extracted ngrok.exe on your local machine. Once you are there, run ngrok. ngrok allows you to generate a host name using which you are able to access your machine over the Internet. You will need to tell ngrok which port you are intended to expose, which in this case will be 3000.

After running ngrok, simply execute the following command to expose your machine over the Internet:

```
ngrok http 3000
```

Refer to Figure 10-8.

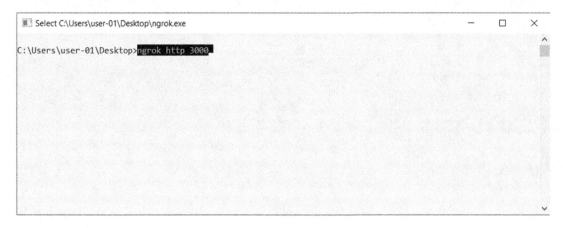

Figure 10-8. *Run ngrok.exe*

Once done, you will have to copy the https instance from ngrok and replace it with your `http://localhost:3000`. Refer to Figure 10-9 to check what exactly you need to copy.

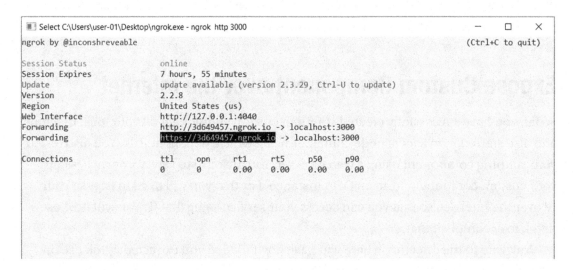

Figure 10-9. *ngrok https host*

Considering the preceding figure, your custom component will be accessible over `https://3d649457.ngrok.io/components`. You may check this by accessing this URL from your web browser. Check Figure 10-10.

Figure 10-10. *Access Custom Component using public URL*

Note   Host which you generated using ngrok will get changed each time you close the ngrok application or if your session gets expired. An expired host will no longer be accessible. However, a permanent host name could be created with a free subscription to ngrok.

Access Custom Component from ODA Instance

Now that you have a custom component running locally and accessible over the Internet, let's see how you can access the same from your ODA instance.

Login to your ODA instance and navigate to the skill in which you want to consume your component. In this case, we want to access this custom component from FindTrip skill (Figure 10-11) which you have created in Chapter 5 of this book.

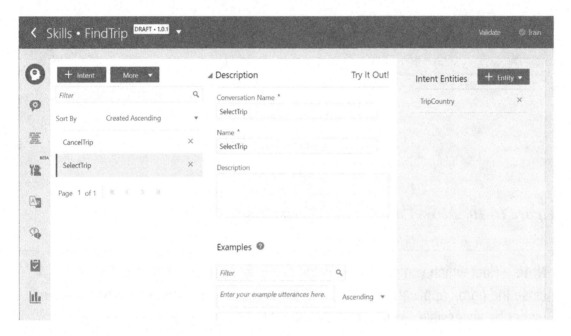

Figure 10-11. *FindTrip Skill*

From left-hand menu items, click 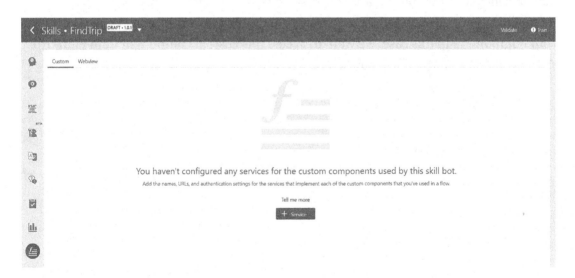 Components. You create custom component on "Custom" tab of the screen. As you have not yet associated any custom component service to the skill, the screen looks as shown in Figure 10-12.

Figure 10-12. *ODA components screen*

Click the **+ Service** button to create a new custom component service. This will give you a new Window as shown in Figure 10-13.

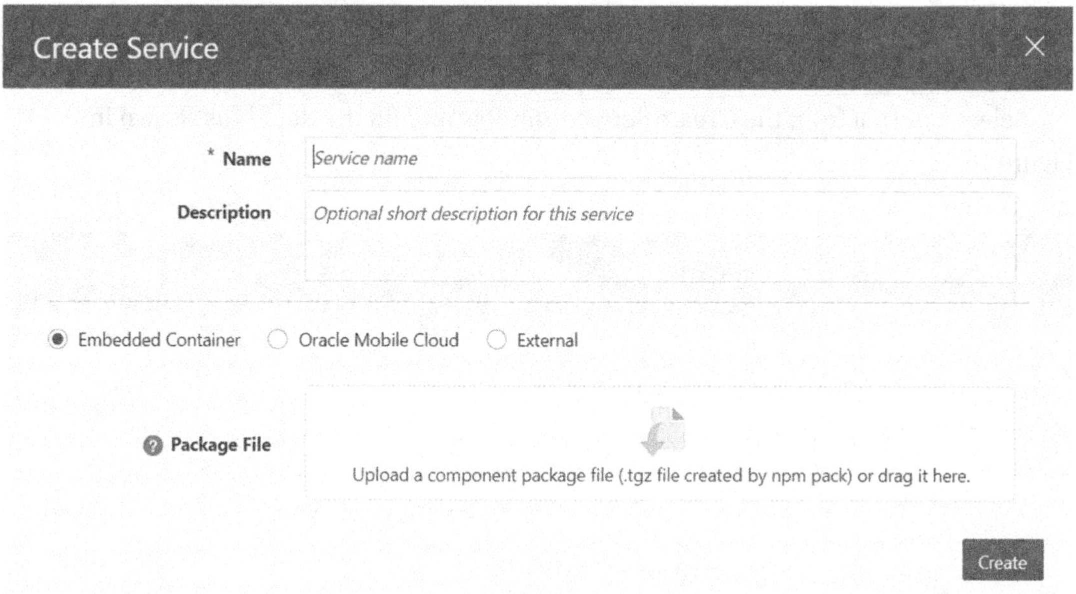

Create Service   ×

* **Name** | Service name

Description | Optional short description for this service

◉ Embedded Container   ○ Oracle Mobile Cloud   ○ External

❓ **Package File**

Upload a component package file (.tgz file created by npm pack) or drag it here.

Create

Figure 10-13. *Create service*

From Figure 10-13, you can see that you can create custom component service in three different ways:

- Embedded Container: You will use this approach when you want to deploy your custom component in a node container embedded inside your ODA instance. This approach will be explained in later section of this chapter. You will use this approach once your custom component will be ready.

- Oracle Mobile Cloud: This approach can be used if you want to expose your custom component as an API service. For this approach, you need to set up an Oracle Mobile Hub.

- External: You will use this approach if you want to use a custom
  component deployed and running on an external node container.
  This approach suites our case as we already have the component up
  and running in our local environment. As an alternative, you can also
  use any Node-capable server. Where you can create a node platform
  and host your custom component from.

Select External from the Create Service window and fill the details as shown in
Figure 10-14.

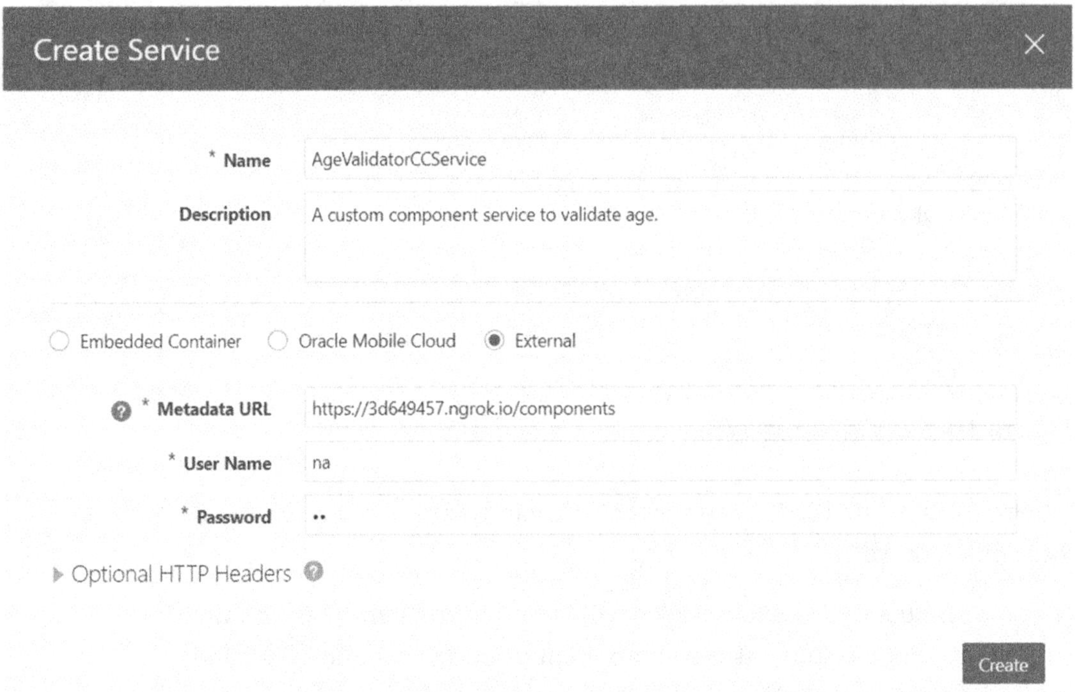

Create Service   ✕

* Name	AgeValidatorCCService
Description	A custom component service to validate age.

○ Embedded Container   ○ Oracle Mobile Cloud   ● External

❓ * Metadata URL	https://3d649457.ngrok.io/components
* User Name	na
* Password	••

▶ Optional HTTP Headers ❓

Create

Figure 10-14. *Custom Component external service*

User Name and Password are not required for debugging purpose. But since these
are mandatory fields in configuration, you need to provide them.

Once done, finalize the process by clicking Create button. This may take a few

moment based on node modules being used in your custom component, which you
define as dependencies in your package.json. Once done, you will see a new screen
showing details of your custom component. Refer to Figure 10-15.

Figure 10-15. *Custom Component service created*

Now you can see that your custom component service is accessible from ODA instance and is ready to be consumed in FindTrip skill dialog flow.

In upcoming sections, you will update the default implementation of your component with your code to validate age, package the custom component, and deploy to the Embedded Container. Finally, you will invoke the custom component from your bot dialog flow.

Update Custom Component Implementation and Local Debugging

In this section, you will implement the logic for your age validation and debug your component locally for any issue. Copy the following code snippet to your AgeValidator.js:

```
'use strict';

module.exports = {
  metadata: () => ({
    name: 'com.hiking.AgeValidator',
    supportedActions: ['allowed', 'denied']
  }),
```

```
  invoke: (conversation, done) => {
    // perform conversation tasks.
    const userInput = conversation.text();
    // determine age
    // const minBookingAge = 18;
    let age = 0;
    if (userInput){
      const matches = userInput.match(/\d+/);
      if (matches) {
          age = matches[0];
      } else {
          conversation.invalidUserInput("Age input not understood. Please
          try again");
          done();
          return;
      }
    } else {
      var errText = "No age input provided";
      conversation.logger().error(errText);
      done(new Error(errText));
      return;
    }
    conversation.logger().info('AgeValidator: age entered by user=' + age);
    conversation.transition( age >= minBookingAge ? 'allowed' : 'denied' );

    done();
  }
};
```

Let's have a brief overview of these changes. As part of the preceding change, you have updated both metadata and invoke functions from default implementation. In metadata function, you have updated the component name from "AgeValidator" to "com.hiking.AgeValidator". This gives your component a unique name. Though this is not a mandatory step, by doing so you, will be able to categorize your component packages and components which are contained in those packages. This will further help you to reduce the risk of naming clashes with other custom components that may be added by the other bot designers. This way of naming custom components will come much handy as your custom component landscape will grow.

Then you have removed the properties object (which was earlier part of default implementation) as you don't need that here and also updated the supportedActions. Later in this section, you will notice how these actions will be used for state navigation in dialog flow.

Next, you updated invoke function to get user's response using conversation.text() method and stored it in a variable. Using regular expressions, you then extracted the numeric value of age from the variable in which you stored user response. After that you have added conditions to validate the age and response for those conditions. Take a closer look at various conversation methods which are being used as part of this implementation.

Main benefit of local debugging is that it allows you to step through the code and find issues, in case you have.

Alright, now that your implementation is ready, save the file. Then restart your custom component ageValidatorCC by navigating to parent directory, which in this case is ODACustomComponents. Use the following command, as explained previously in the "Run Your Custom Component Locally" section:

//use ctrl+c in the existing command window to stop the existing process

```
bots-node-sdk service ageValidatorCC
```

Note   Every time you make a change in your local component and save it, you need to restart your local node container using the preceding command.

From your local machine if you now access http://localhost:3000 from a browser, you will notice that these new changes from AgeValidator.js are being reflected. Check Figure 10-16.

{"version":"1.1","components":[{"name":"com.hiking.AgeValidator","supportedActions":["allowed","denied"]}]}

Figure 10-16. *Update component response*

If your ngrok session is still active, you just need to reload your custom component in ODA instance. To do so, you need to click the [Reload] button available in the top-right corner of the Components screen in ODA.

If your ngrok session is expired or terminated, you need to regenerate a host name using ngrok as explained in the "Expose Custom Component over the Internet" section. After generating a new host name for port 3000, you need to update the Metadata URL on the Components screen. Refer to the highlighted section in Figure 10-17.

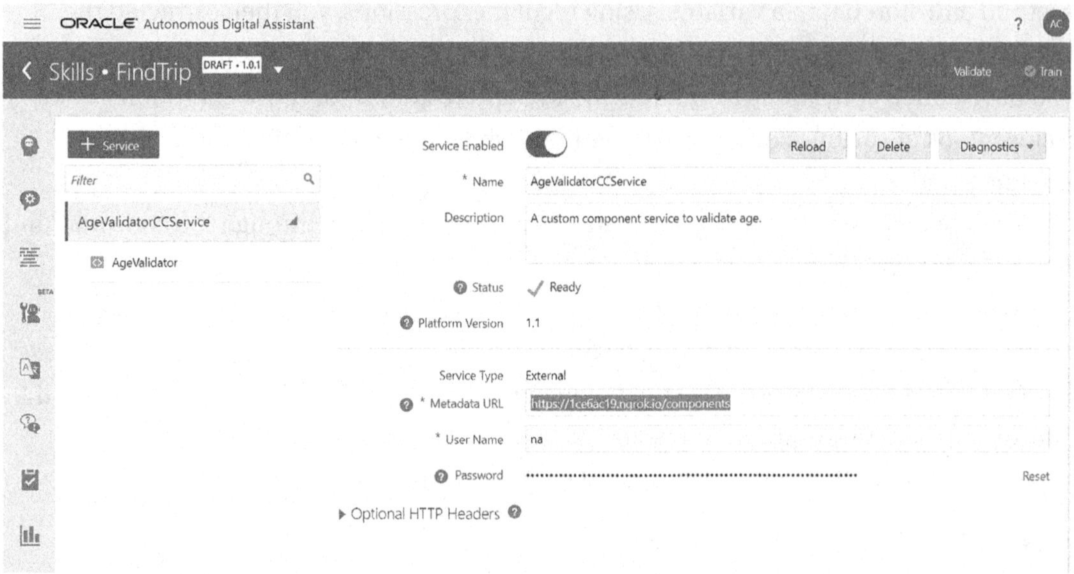

Figure 10-17. *Update Metadata URL post generating new host*

Presuming that your custom component is running locally and is accessible in ODA as explained in the preceding text, now is the time to invoke the custom component from your bot's dialog flow. Navigate to [icon] Flows from the left-hand menu. Once you are on the Flows screen, update the dialog flow as follows:

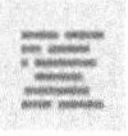

```
metadata:
  platformVersion: 1.0.1
main: true
name: FindTrip
context:
  variables:
```

```
states:
  intent:
    component: "System.Intent"
    properties:
      variable: "iResult"
    transitions:
      actions:
        SelectTrip: "askAge"
        CancelTrip: "cancelTrip"
        unresolvedIntent: "unresolved"

  askAge:
    component: "System.Output"
    properties:
      text: "How old are you?"
    transitions:
      next: "validateAge"

  validateAge:
    component: "com.hiking.AgeValidator"
    properties:
    transitions:
      next: "bookTrip"
      actions:
        denied: "underAge"

  bookTrip:
    component: "System.Output"
    properties:
      text: "You will be soon able to book trip using Travvy"
    transitions:
      return: "done"

  underAge:
    component: "System.Output"
    properties:
      text: "You don't meet minimum age criteria to book a trip using Travvy."
```

```
    transitions:
      return: "done"

  cancelTrip:
    component: "System.Output"
    properties:
      text: "You will be soon able to cancel your trip using Travvy"
    transitions:
      return: "done"

  unresolved:
    component: "System.Output"
    properties:
      text: "Sorry I don't understand you. Please try again"
    transitions:
      return: "done"
```

As part of the preceding bot flow, you first used "System.Intent" component to identify intent. These are the same intents which you created as part of Chapter 5. Based on intent resolution by System.Intent component, you have then defined the state corresponding to those intents. Notice that when an intent gets resolved to SelectTrip, only then you are doing an age validation else not.

Check "askAge" state where you are asking user to provide his/her age and then making a transition to "validateAge" state. This state is invoking your custom component. Then based on the response of custom component, you are passing bot handle to different states.

As the skill implementation is in its early stages, you will only display static text content for various states. There is nothing wrong in that, and it also suffices the purpose of custom component implementation. These texts can be easily updated to more appropriate bot response as your bot evolves.

Now that you are having a clear understanding of the changes you made in custom component implementation and bot flow, let's test the skill using Skill Tester. From the left-hand menu, click ▶ to open Skill Tester. Write "I want to book a trip" in the message section, refer to Figure 10-18, and press Enter.

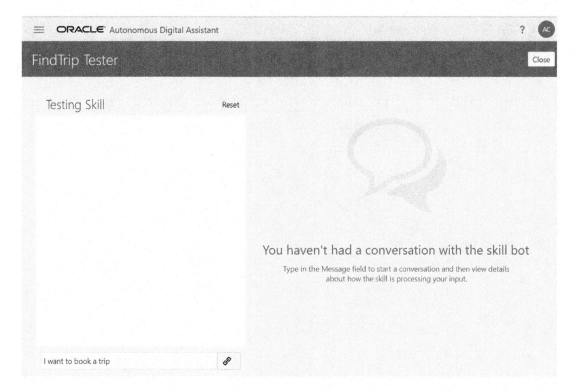

Figure 10-18. *Skill Tester*

What happened when you enter your age? Did you get an error message in Skill Tester as shown in Figure 10-19?

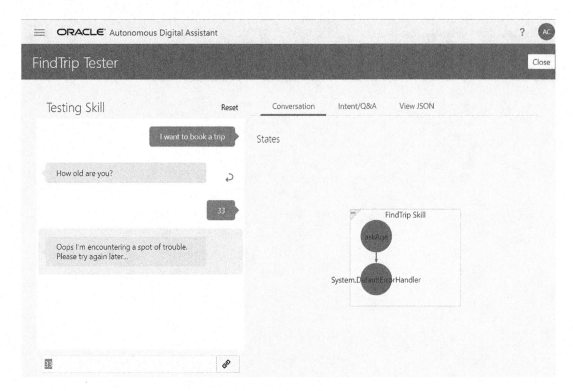

Figure 10-19. *Error message*

You didn't expect this, right? Never mind, this gives you an opportunity to debug your code.

Because of the local setup of custom component development, which you did in previous sections of this chapter, it becomes much easier to debug issues and identify the error. Let's check if there is any error in command prompt of your local machine. Did you see any error? There is one. Refer to Figure 10-20.

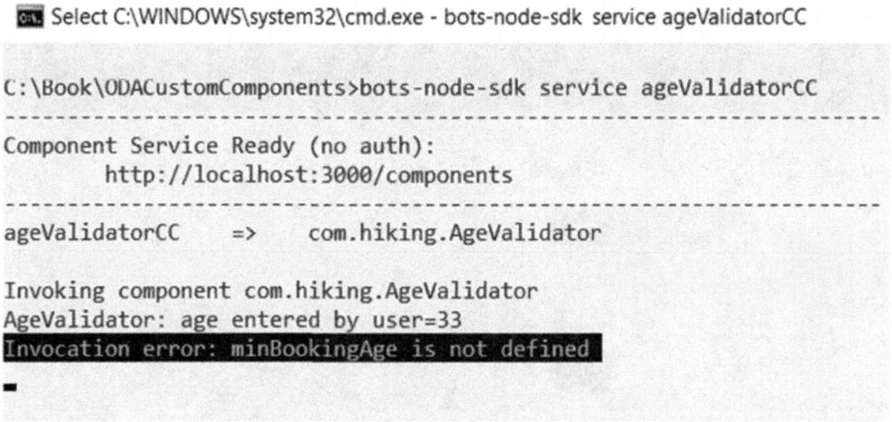

Figure 10-20. *Custom Component error*

We intentionally introduced this error during the course of implementing your AgeValidator.js logic to demonstrate how you can debug your code locally.

Check your AgeValidator.js and uncomment the line where you have defined the variable "minBookingAge". The following is the code change:

From:

```
// const minBookingAge = 18;
```

To:

```
const minBookingAge = 18;
```

Once updated, restart the component service. Reset your Skill Tester by clicking the

Reset button and test the same scenario. It will be successful this time, and you will

receive response as shown in Figure 10-21.

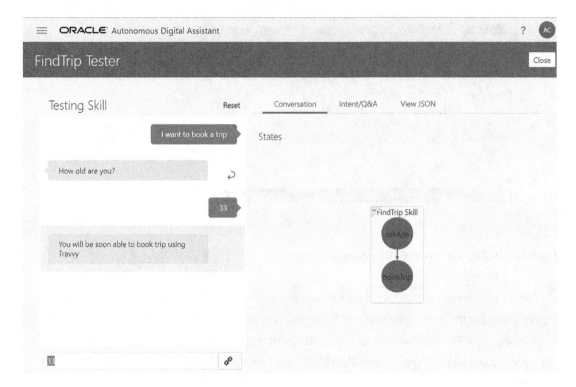

Figure 10-21. *Successful test execution*

In this section, you have successfully implemented your custom component and invoked that component from FindTrip skill. You have also implemented your bot flow to invoke that custom component.

You might also want to check a similar custom component implementation in PizzaBot which is one of the sample skills provided to you by Oracle as part of ODA instances. Where the skill checks your age before allowing you to order pizza. The custom component implementation which you did in this section is also influenced by the same concept.

Now that your custom component is ready in your local environment and that you have also tested it, this brings you to the final section of this chapter, where you will see how you can deploy your custom component to an Embedded Container in ODA instance. Let's check that.

Deploying a Custom Component to Embedded Container in ODA Instance

To deploy a custom component to Embedded Container in ODA instance, you first need to package your component. There will be absolutely no change in terms of your custom component implementation or in the dialog flow of your skill.

Open the command prompt and navigate to ageValidatorCC directory on your system where you have created your custom component implementation. From there, execute the following command:

```
npm pack
```

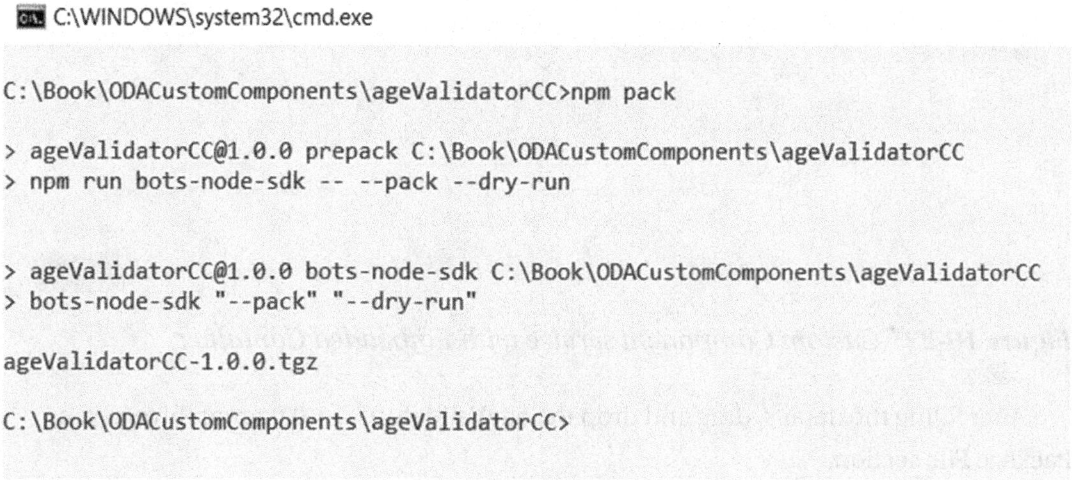

```
C:\WINDOWS\system32\cmd.exe

C:\Book\ODACustomComponents\ageValidatorCC>npm pack

> ageValidatorCC@1.0.0 prepack C:\Book\ODACustomComponents\ageValidatorCC
> npm run bots-node-sdk -- --pack --dry-run


> ageValidatorCC@1.0.0 bots-node-sdk C:\Book\ODACustomComponents\ageValidatorCC
> bots-node-sdk "--pack" "--dry-run"

ageValidatorCC-1.0.0.tgz

C:\Book\ODACustomComponents\ageValidatorCC>
```

Figure 10-22. *Packaging Custom Component*

As you can see in Figure 10-22, this generated an archive with name ageValidatorCC--1.0.0.tgz inside ageValidatorCC directory.

Next, navigate to the Components screen by clicking the ⬚ button from the

left-hand menu.

From Components screen, you first need to delete the existing deployment where you are referring to the component running on your local machine. To do so, click the 
Delete button which you will find in the top-right corner of your screen. When you click the button, you will see a new window asking you to provide the confirmation to delete the component. Agree to that.

Now you should have an empty Components screen with [+ Service] button in the middle of the screen. You will once again start the same process of creating a new component service. Click the Service button and this time fill in the details for Embedded Container as shown in Figure 10-23.

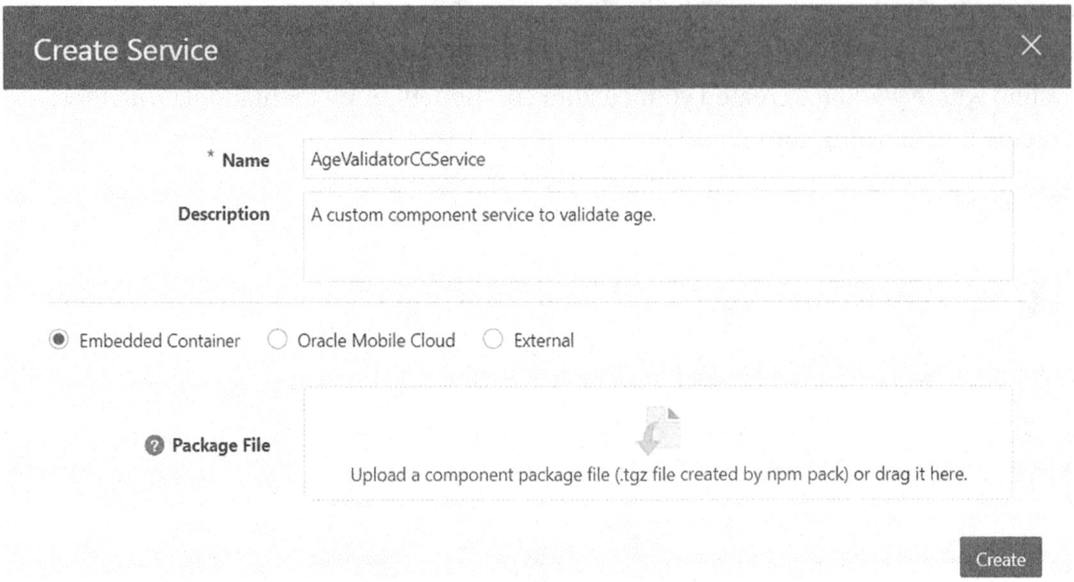

Create Service   ✕

* **Name**	AgeValidatorCCService
Description	A custom component service to validate age.

◉ Embedded Container   ○ Oracle Mobile Cloud   ○ External

❔ Package File

Upload a component package file (.tgz file created by npm pack) or drag it here.

[Create]

Figure 10-23. *Custom Component service with Embedded Container*

After filling the details, drag and drop the ageValidatorCC-1.0.0.tgz archive in the Package File section.

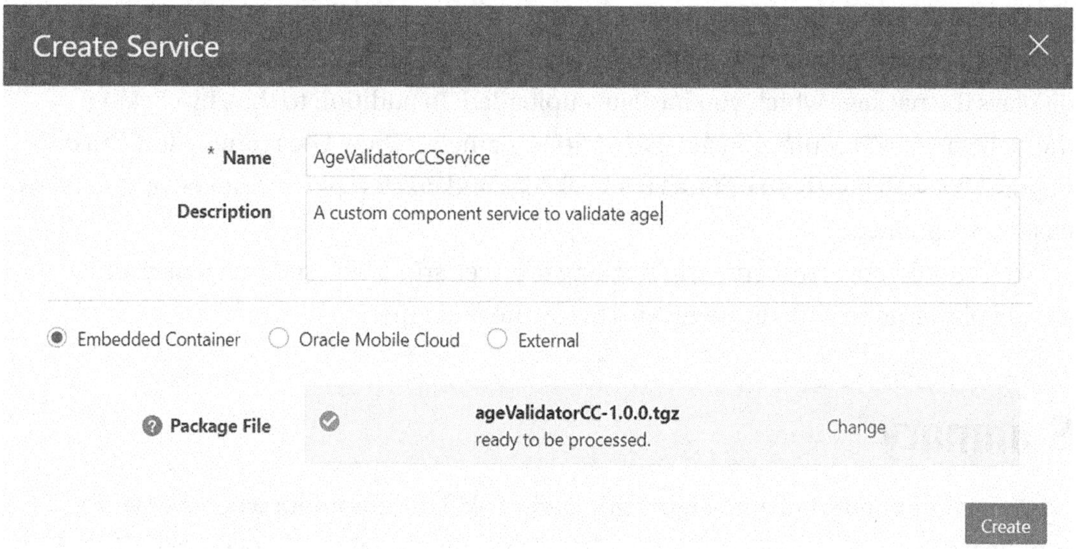

Figure 10-24. *Custom Component upload*

Click the Create button as shown in Figure 10-24 to create the service. Wait for a while till the deployment gets finished. Once done, you should get a screen as shown in Figure 10-25.

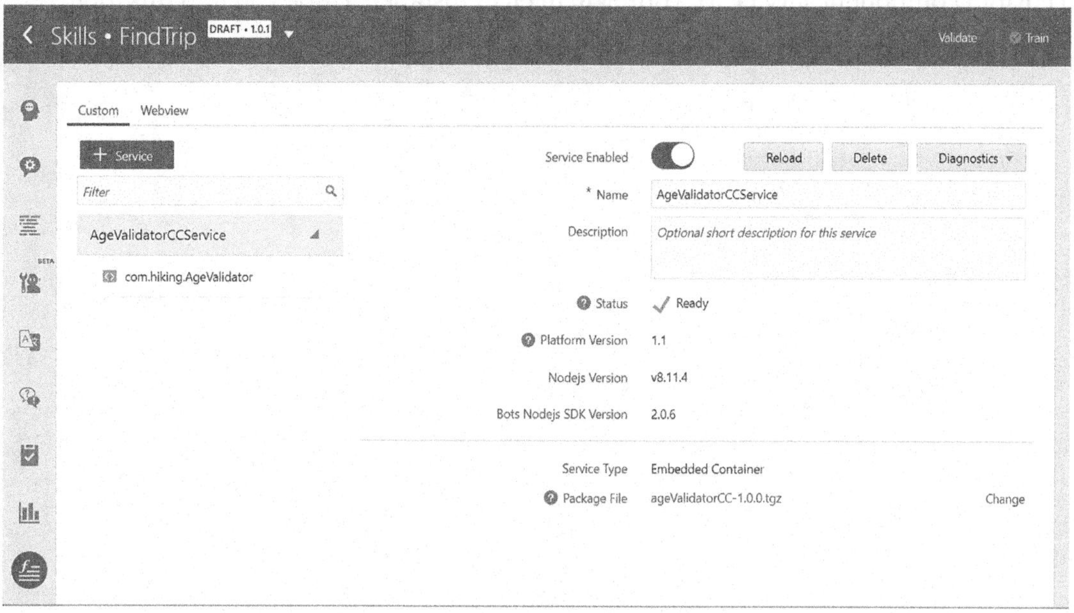

Figure 10-25. *Custom Component deployed on Embedded Container*

Notice that all details related to your component remain the same as earlier except this time you can see that the "Service Type" is "Embedded Container" which displays the package which you have just uploaded. In addition to that, by clicking the "Diagnostics" button, you get two options, namely, "View Logs" and "View Crash Report." You can use these options to check logs and crash reports, respectively, for your custom component.

You should now reexecute trip booking test scenario discussed previously and should play around with different ages to see the outcome.

Summary

Custom components become an integral part of Skill development whenever you want to implement a functionality which built-in components cannot support. In the beginning of this chapter, you were introduced to custom components, and by the time you reached the end, you are ready with a custom component deployed and running in Embedded Container. By now, you must have realized how important it is to have a local development environment for custom components to be able to create, run, and debug components on your machine. You are now also aware of different approaches to create a custom component service for your Skill in ODA instance. Hope you find this topic interesting, and will see you again in the next chapter.

CHAPTER 11

Adding Speech and Sentiment to Your Digital Assistant

Introduction

In the previous chapters of this book, you learned how to build some very smart Digital Assistants based on Oracle technology. They work just fine over many different channels and cover most of the use cases that you can think of. In this chapter, you will learn how to add some extra dimensions to your Digital Assistant. First, you will learn how to enrich a Digital Assistant with sentiment analysis, and next you will learn how to add speech to your Digital Assistant. While reading this chapter, it will become clear to you that both additions are based on a sophisticated usage of the knowledge that you already acquired in the previous chapters of this book.

Sentiment Analysis

If you want to create a chatbot that behaves more like a human than like a robot, the chatbot should be able to understand what a user is saying and, even better, have an understanding of context, tone, and even subtle nuances like sarcasm. The most obvious way of achieving such behavior is the introduction sentiment analysis.

According to the dictionary, sentiment analysis is *"the process of computationally identifying and categorizing opinions expressed in a piece of text, especially in order to determine whether the writer's attitude towards a particular topic, product, etc. is positive, negative, or neutral."*

© Luc Bors, Ardhendu Samajdwer, Mascha van Oosterhout 2020
L. Bors et al., *Oracle Digital Assistant*, https://doi.org/10.1007/978-1-4842-5422-6_11

When you add sentiment analysis to your bot, it will be able to better understand the user. If the bot can sense whether the user is pleased, upset, or angry, it will be able to resolve intents better and most probably make the skill react less like a bot and behave more like a human. This adds to user satisfaction which in turn can have positive effects on your business. This is all based on human psychology. People who feel happy or neutral tend to take bad news or frustration in a more accepting way, whereas sad, disappointed, or mad people will have very limited patience.

To make this work, you must be very careful. In the past, experiments such as with Microsofts' Tay chatbot have failed. Tay became homophobic, racist, and Nazi, just by interacting with users. Tay learned based on the users' responses, and the users in turn were teaching Tay to become a "bad bot." You can easily understand that this should not happen to your companies' bots. Things like syntax, sarcasm and cynicism, choice of word, cultural differences, and any other factor, which to a human can be very normal, have to be taken into consideration and can influence your bot's response. There are no simple solutions to this. So whenever you decide to add sentiment analyses to your bot, you must make sure not to damage the perception and reputation of your brand.

In order to add this kind of functionality to an Oracle Digital Assistant, you could use a third-party service as there is no built-in solution in Oracle Digital Assistant.

What to Consider Before Implementation?

Sentiment analysis can be implemented in various ways in the context of a chatbot, but the implementation should be always influenced by your requirement. Invoking any third-party API, for a commercial purpose, will involve a cost. Hence, it becomes crucial to understand the scope and the available budget before you start the implementation.

Having these facts clear, let's consider a few real-life scenarios. Miscreant users often try to break a bot flow by asking irrelevant questions. for example, "are you male or female?", "how is the weather?", "are you a bot?", "do you love me?", and so on. You should try to break any such conversation at the beginning itself and try to bring back the user to your bot's context. The most common approach to handle any such conversations can be to add them to a skill's "unresolvedIntent." Or else, you can also create a specific intent in your skill to handle such conversations. This could be a dedicated "small talk" skill, which allows the Digital Assistant to cope with such questions and keep the conversation going. Keep an eye on such conversations using your skill's Insight and keep adding those user inputs either to "unresolvedIntent" or in your specific intent.

Another probable way, if you have exposed skill via a Digital Assistant and not the Skill directly over a channel, can be to add an explicit Skill to handle such chit-chat conversations. If you wish to implement this approach, then you don't need to add abovementioned user inputs to any specific intent or "unresolvedIntent." Rather, you create intent(s) in your dedicated skill to handle such conversation and let your Digital Assistant to take care of routing. This approach allows you to the keep the logic behind your business-specific skills clean.

If the abovementioned approaches are not what exactly you are looking for and rather you want to rely on a more sophisticated approach, then using third-party APIs is the way to go forward. As mentioned before, use of third-party APIs will always adhere to a cost, but they also come with various benefits. The foremost is the fact that while using such API services, you don't have to implement any complex feature to analyze user sentiment within your skill or digital assistant. As these APIs are built for highly commercial purposes, you can rely on their precision of analyzing the user input.

Such APIs analyze the user inputs in broadly three categories, that is, positive, neutral, and negative. But there are services which further provide you additional categories such as extremely positive or very positive and extremely negative or very negative. While using such services, based on the analysis made by them on user's input, you can then design appropriate dialog flow of your bot. For example, consider a case where you send a feedback form to the end user based on API result.

There are various commercial APIs available today for sentiment analysis. In the following, you can find the list of a few popular sentiment analysis APIs:

- Google Sentiment Analysis

- IBM Watson Tone Analyzer API

- Qemotion

- AYLIEN API

- PreCeive API

- MoodPatrol API

- Indico API

- ParallelDots

- DeepAI

In order to use any of the sentiment analysis API of your choice, you will first have to register with the API provider on their web site. This will allow you to get the access token. Using the token, you will then be able to invoke the API from your chatbot. This process should be quite simple, and as you can choose any of the available API providers, this depends entirely on you. Typically these third parties provide a key or token during setup that should be used when invoking their API. We would suggest keeping the access token handy with you as it will be required in upcoming sections.

Analyzing Sentiment from Text

In order to analyze text using any third-party API service, you simply pass the user's input from your skill to the API. API then analyzes the input value and returns you the outcome. Based on the API outcome, you can then design your chatbot flow.

To invoke the sentiment analysis API from your skill, you will be using a custom component. Custom components have already been discussed in detail during the course of Chapter 10.

The process will be straight forward. From your skill, you will have to pass the user input to the custom component at the time of custom component invocation. Inside your custom component, you will then extract the user input from **conversation** argument of **invoke** function (conversation is a reference to bots-node-sdk). Then you would pass the extracted input to the sentiment analysis API. Invocation call to the sentiment analysis API may vary from provider to provider, which you need to take care of. Once you receive the API result, pass it back to the skill from your custom component. That's all what you need to do. Each of these steps will be further described in detail in next section.

Invocation of Sentiment Analysis API

Create a skill in your Oracle Digital Assistant environment and name it as TravvySentimentAnalysis. For the time being, this skill will be implemented with bare minimum functionality for the sentiment analysis demonstration purpose. Add the following code in Flows:

```
#metadata: information about the flow
#  platformVersion: the version of the bots platform that this flow was
written to work with
metadata:
```

```
  platformVersion: "1.0"
main: true
name: TravvySentimentAnalysis
#context: Define the variables which will used throughout the dialog flow
here.
context:
  variables:

#states is where you can define the various states within your flow.

states:
  askFeedback:
    component: "System.Output"
    properties:
      text: "How was your experience with Travvy?"
    transitions:
      next: "checkFeedback"

  checkFeedback:
    component: "com.hiking.SentimentAnalysis"
    properties:
    transitions:
      actions:
        Positive: "Positive"
        Negative: "Negative"
        Neutral: "Neutral"

  Positive:
    component: "System.Output"
    properties:
      text: "Thanks for your positive feedback"
    transitions:
      return: "done"

  Negative:
    component: "System.Output"
    properties:
```

```
        text: "Thanks for your response. Apologies for a bad experience"
      transitions:
        return: "done"

  Neutral:
    component: "System.Output"
    properties:
      text: "Thanks!"
    transitions:
      return: "done"
```

As part of first state **askFeedback**, you will request the user to provide feedback of Travvy. Once the user responds back, you will then invoke the custom component **com. hiking.SentimentAnalysis** in the state **checkFeedback**. In subsequent steps, state transition will be performed based on the action response of the custom component.

Next, create a custom component with the name **sentimentAnalysisCC** following the steps mentioned to you during the course of Chapter 10. Once done, create a JavaScript file with the name **SentimentAnalysis.js** inside your custom component. This has also been explained in the same Chapter 10. After that add the following code in the SentimentAnalysis.js:

```
var request = require('request');

module.exports = {
  metadata: () => ({
    name: 'com.hiking.SentimentAnalysis',
    supportedActions: ['Positive','Negative','Neutral']
  }),
  invoke: (conversation, done) => {
    // perform conversation tasks.
    // retrieve user input and store it into a variable
    const userInput = conversation.text();

    // Invoke API
    request.post({
      url: 'https://api.deepai.org/api/sentiment-analysis',
      json: true,
      headers: {
```

```
      'Api-Key' : 'YourAPIKeyGoesHere',
    },
    form: {
      'text': userInput,
    }
  }, function(error, response, body){
    if (response.statusCode === 200) {
      if (body.output[0] === 'Positive') {
        conversation.transition('Positive');
        done();
        return;
      } else if (body.output[0] === 'Negative') {
        conversation.transition('Negative');
        done();
        return;
      } else if (body.output[0] === 'Neutral') {
        conversation.transition('Neutral');
        done();
        return;
      }
    } else {
      conversation.logger().error(errText);
      conversation.transition();

      done();
      return;
    }
  });

  }
};
```

Inline comments have already been added for the steps described in the preceding code snippet. For this case, we have used the sentiment analysis API of DeepAI. After updating your custom component with the preceding changes, all you need to do is to package and deploy. Again, packaging and deployment of custom components have been described in detail in Chapter 10.

Once you complete these steps, you should try to test this skill. You may use in-built Skill Tester for this purpose. Figure 11-1 displays a simple test case of what you have done so far in this section.

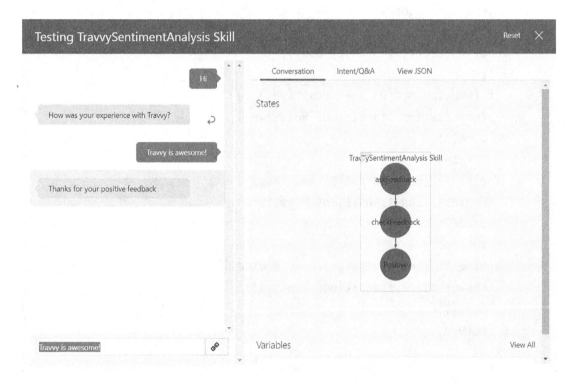

Figure 11-1. *Sentiment analysis from text*

Speech

Nowadays, there's a growing use of speech-enabled personal assistants, such as Google Home, Amazon Alexa, Apple Siri, Microsoft Cortana, and many more. What if you could add speech to Oracle Digital Assistant? As you have already learned, Digital Assistant can be exposed over several predefined channels. Other channels that are not supported out of the box can also be used, but in order to do that, you must use an HTTP **webhook channel** and manually integrate the Digital Assistant with that channel.

The key question is: Can speech be considered a separate channel, and if so, how can this be configured and implemented in Oracle Digital Assistant? Obviously, the answer to the first part of the question is yes. Speech can be considered a separate channel, and to be more precise, you would have to define a separate channel for Amazon Alexa, Google Home, and others.

With that out of the way, you will now learn how to implement speech for Oracle Digital Assistant.

Note   At the moment of the writing of this book, Amazon Alexa, Google Home, and other third-party provider services were the only way of exposing Oracle Digital Assistant over speech channels. Oracle recently acquired speak.ai and is working on implementing their own speech engine and adding it to Oracle Digital Assistant. When that is in place, you can still use the third-party solutions, but Oracle then provides you with the option to enable speech interaction completely embedded in the Oracle Digital Assistant without having to rely on third-party services.

What Do You Need?

Before you learn how to implement the webhook channel, you first need some understanding of what a webhook is and how it works. In short webhooks, technically, are "user-defined callbacks made with HTTP." Webhooks are typically used to connect two different applications. When an event happens on the trigger application, it sends (POST) data to a webhook URL from the action application. Webhooks use "event reaction" (don't call us, we'll call you). This concept avoids the need for constant polling from the receiving end to the sending end.

To create a webhook channel in Oracle Digital Assistant, you need the following:

- A publicly accessible HTTP messaging server that relays messages between the user device and your bot using a webhook. Typically, in an Oracle Cloud environment, you would use an Oracle Compute Instance to host this, but any node server will do. You implement this webhook in node, and it will need to implement the following two endpoints:

    - A POST to receive messages from our bot

    - A POST to send messages to our bot

- The URI of the webhook call that receives your bot's messages (so the bot knows where to send the messages).

- The webhook URL that's generated for your bot after you complete the Create Channel dialog (so that the message server can access your bot).

For all these pieces to come together in a working webhook-based solution, you need to set up the node server, configure the webhook channel on Digital Assistant, write the Node.js implementation for the incoming and outgoing webhook, and configure the speech engine(s).

Note   For this purpose of explaining how to add speech to Oracle Digital Assistant, Amazon Alexa is used. Other platforms can be added in the same way such as Google Assistant, as long as they support webhooks.

Figure 11-2 shows the overall picture of this solution:

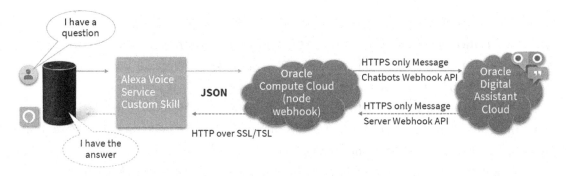

Figure 11-2.   *Digital Assistant to Alexa architecture*

Note   Extended documentation of the Alexa app node package can be found at www.npmjs.com/package/alexa-app.

You can even change the way Alexa is speaking by using an ssml-builder to manipulate the output. The "**ssml-builder /amazon_speech**" node package can add Alexa-specific SSML tags to the output, thus instructing Alexa to speak in a different way.

Note   Extended documentation of the ssml-builder node package can be found at www.npmjs.com/package/ssml-builder.

Speech Synthesis Markup Language (SSML)

Speech Synthesis Markup Language (SSML) is an XML-based markup language. SSML allows developers to specify how input text is converted into synthesized speech using a text-to-speech service. SSML enables developers to fine-tune the pitch, pronunciation, speaking rate, volume, and more of the text-to-speech output. Punctuation, such as pausing after a period or using the correct intonation when a sentence ends with a question mark, is automatically handled.

The World Wide Web Consortium has a recommendation regarding Speech Synthesis Markup Language, which can be found here `www.w3.org/TR/speech-synthesis11/`.

With SSML you can use specific text to instruct text-to-speech services how they should "speak" your text, as can be seen from the code sample as follows:

```
"<speak>Hi, this output speech uses SSML. It is part of this chapter.</speak>"
```

This example could be enhanced by adding extra tag. The next example shows how to add a 1 second pause to the sentence:

```
"<speak>Hi, <break time="1" /> this output speech uses SSML. It is part of
this chapter.</speak>"
```

You can also change the voice that is used to speak a part of your sentence:

```
"<speak>Hi, <break time="1" /> this output speech uses SSML. <voice
name="Kendra"> It is part of this chapter. </voice></speak>"
```

And there are many more ways to change to text-to-speech behavior.

Note   SSML is not fully standardized (yet). Any vendors use their own specific implementations and additions to their platform. Make sure that you know what tags are supported by the platform that you are targeting.

The Webhook Code

To implement the webhook code, you have two options. The first one is to entirely code all logic and plumbing yourself. This might give you some flexibility but also adds a lot of coding to your task. The second option, which we also used for the purpose of this book, is probably a bit smarter. The Oracle Product Management team created a working example library that can be used as a basis for your Alexa integration. This code uses the bots-node-sdk. The bots-node-sdk contains the following libraries to facilitate communication between your webhook and the Oracle Digital Assistant:

- OracleBot.Util.Webhook

- OracleBot.Util.MessageModel

- OracleBot.Lib.MessageModel

- OracleBot.Util.Text

So instead of coding the plumbing such as message signing and modeling the messages between Alexa and ODA, you can simply use the bots-node-sdk via a require definition and from there use the libraries that are inside that SDK.

```
const OracleBot = require('@oracle/bots-node-sdk');
const MessageModel = OracleBot.Lib.MessageModel;
const messageModelUtil = OracleBot.Util.MessageModel;
const botUtil = OracleBot.Util.Text;
const webhookUtil = OracleBot.Util.Webhook;
```

With this generic piece of node code in place, you can use the webhook with any skill that you want to expose over Alexa. The final configuration of the webhook between Alexa and ODA also needs to be done in this node code. You will learn later in this chapter how to do that.

Looking at Travvy

The process of exposing Travvy over Alexa is rather straightforward and consists only a couple of configuration steps. You will be guided through these steps, and they will be explained individually.

Adding the Webhook Channel to ODA

The first step to create a new channel in ODA. This has to be a webhook channel as this is the type of channel that is required by this kind of channel. You can simply invoke the "**create channel**" button and fill out all the fields that you need (Figure 11-3). Note that in the "Outgoing webhook URI," you must set your server URL plus the endpoint where the webhook code is expecting the messages from ODA to go. In this case **https://your-server-url/genericBotWebhook/messages** is used because the URL where the webhook code will run is not yet known. The endpoint will be **/genericBotWebhook/Messages** as this is how it is defined it in the webhook code.

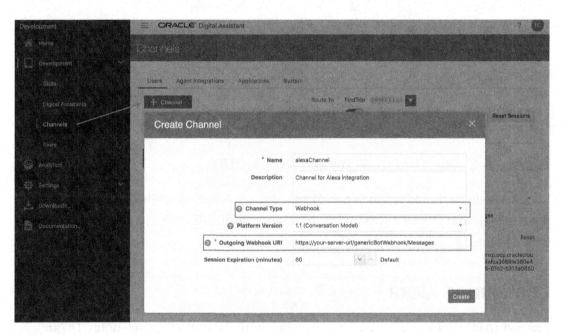

Figure 11-3. *Create channel in ODA*

When you click OK, the webhook channel will be created for you, and you will see (Figure 11-4) two extra fields added, both of which you need in the configuration of the webhook code:

1) The webhook URL: The URL that your webhook will use to send messages to the ODA

2) The secret key: The key that your webhook needs to use to sign messages that are sent to ODA and that ODA will use to check if incoming messages are signed with the correct key

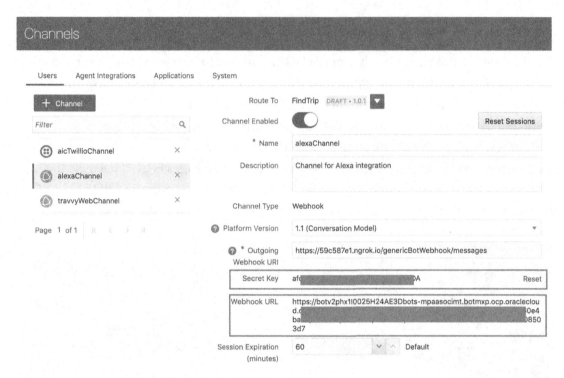

Figure 11-4. *Generated secret key and webhook URL*

This is all that is needed in Oracle Digital Assistant. Just by adding and configuring the channel, your Digital Assistant (or Skill) is now available over this webhook channel which eventually will be used by Alexa.

Configuring Alexa

The way to interact with Alexa is by sending a command (either speech or text) to an Alexa-enabled device. This command will be forwarded to a skill running on Amazon's Alexa Service and eventually to services or data sources that can provide the requested information. To enable Oracle Digital Assistant to work with Alexa, a custom skill has to be created, which will invoke a REST endpoint, will forward the Alexa request to our DA, and will also get the DA's response and send it back to Alexa.

To create this skill for Alexa and make it work, Alexa needs to be configured. The steps involved in this will be explained in the next sections.

An Alexa Invocation Sentence

The invocation sentence is needed to begin an interaction with a particular custom skill (Figure 11-5). To invoke Travvy, you can use, for instance, "Travvy travel."

So whenever the user says "Alexa, Travvy travel," this custom skill will be invoked. Now, to book a trip, the user can say "Alexa, ask Travvy travel to book me a trip."

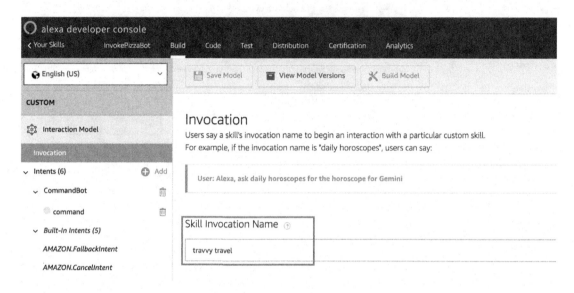

Figure 11-5. *Set up invocation name*

Alexa Intents and Slot Types

With an intent, you try to say what is it that you want, what is your intention. It typically consists of a sentence. Intents can optionally have arguments called slots. Slots are basically *variables in utterances*. A slot can have a predefined value but is by default empty. To define a slot, you first need to create a custom Intent in your Alexa skill (Figure 11-6). After that, when you need to create your sample utterances for the Intent, simply define a slot by writing it in curly brackets.

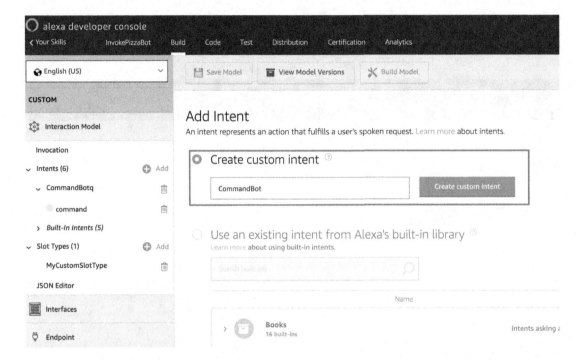

Figure 11-6. *Alexa add intent*

Slot values are extracted from utterances and sent with the intent request. Actual intent resolution will not be done by Amazon Alexa but by our Oracle Digital Assistant. This means that you can use slot type "AMAZON.LITERAL" (Figure 11-7). This slot will hold the text representation of whatever the user said. The entire content of this slot will then be forwarded to the skill's endpoint and to Oracle Digital Assistant for intent resolution.

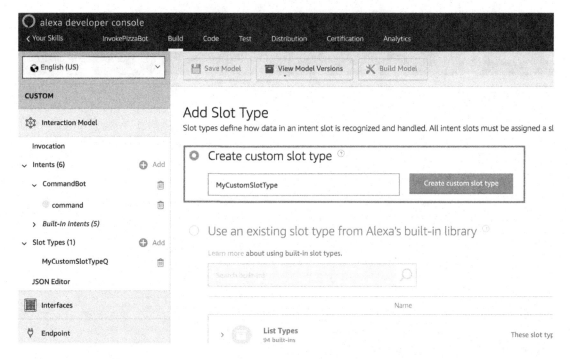

Figure 11-7. *Alexa add a slot type*

An Endpoint

The actual logic for the custom skill will in our case be running on a REST endpoint implemented in Node.js and run on a server that hosts the Alexa app.

For the purpose of this demo, it will be running locally, with an ngrok URL for it. The endpoint needs to be added to root URL of the server that hosts the Alexa app, in our case the local ngrok.

In the HTTPS endpoint field, you need to fill out the exact URL of this (Figure 11-8):

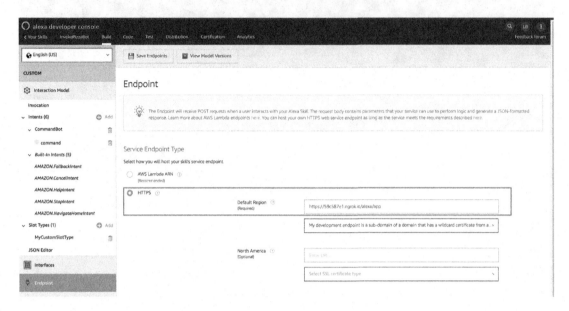

Figure 11-8. *Alexa configure endpoint*

With this in place, the custom Alexa skill is configured. Finally you need to find the Skill ID so it can be used in the configuration section of our node code. The Skill ID can be found from the Alexa Skills Kit (ASK) Developer Console (Figure 11-9):

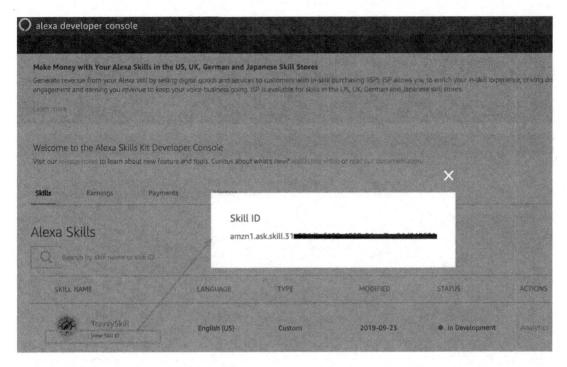

Figure 11-9. *Alexa find the Skill ID*

This ID needs to be added to the node code of the webhook that was previously created so the webhook knows where to find the Alexa skill.

Setting Up the Webhook

In order to set up the webhook between Oracle Digital Assistant and Amazon Alexa, they need to know where to find each other. To achieve this, you need to add the previously generated **Secret Key**, **Channel URL**, and **Amazon Skill ID** to the metadata settings in the JavaScript code:

```
//replace these settings to point to your webhook channel
var metadata = {
  allowConfigUpdate: true,
  waitForMoreResponsesMs: 200,
  amzn_appId: "<Your Amazon Alexa Skill ID>",
  channelSecretKey: '<Your BOT Secret Key>', //BOT Secret Key
  channelUrl: '<Your BOT Channel URL>'
};
```

Testing the Alexa Channel

The conversation over Alexa can be easily tested with the Alexa developer console
(Figure 11-10). Simply go to the test tab, and start typing your commands, or if you like it,
you can even talk to Alexa.

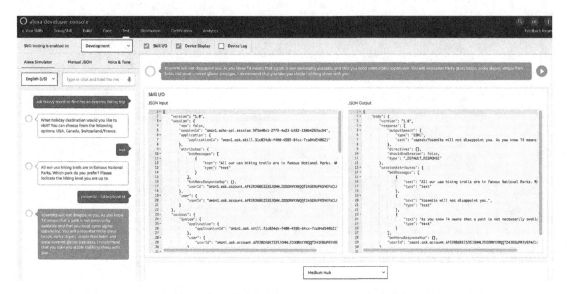

Figure 11-10. *Alexa, testing in the console*

Another option is to use a smartphone which removes the need of involving a real
Amazon Echo or Alexa device. You can use the official Alexa app on your smartphone.
The following example shows how to run Alexa on Android phone. In the Google Play
Store, search for Alexa and select Amazon Alexa. Tap the Install button and wait while
the Amazon Alexa app downloads onto your device. Once you have installed Alexa
on your phone, you will need to set it up. Open the Amazon app and log in using your
existing Amazon account information, including your email address (or phone number,
if you have a mobile account) and password. Tap the Sign In button.

Once logged in, you should be able to use the app as were it an Amazon Echo or Dot
and invoke the skill that calls the Oracle Digital Assistant. Click the skill to view more
information or simply click the Alexa button to talk to Alexa and ask her to get Travvy to
find you a trip (Figure 11-11).

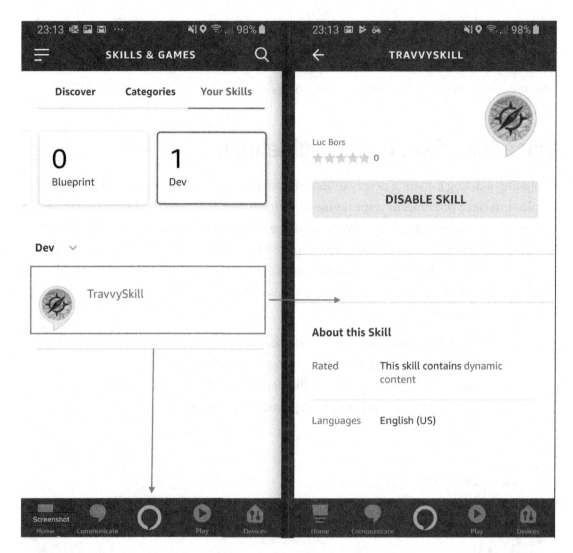

Figure 11-11. *Alexa, skill testing in app*

It is obvious that a conversation over a voice channel is different from a conversation over a text or web channel. A voice channel will not be able to show images, for instance, nor will it allow you to click links or invoke webview components to enter additional data. However, it is possible to work with the differences. The Digital Assistant can send channel-specific replies. In a dialog flow, you can tell about the channel type and the name of the channel being used from expressions and in custom components. This would give you the option to optimize your dialog flow for use with voice.

In addition, the System.CommonResponse component could show or hide response items based on the channel type and name.

In the previous sections, you learned how to technically set up Alexa as a channel. It works, but the conversation is somewhat cumbersome. What you probably notice is that working with a voice channel is that it requires a special design of the conversation. The next section gives you some hints on how to design a voice user experience (VUX).

Voice User Experience Guidelines

Designing a dialog for voice is very different from designing a conversation for vision. A voice interface provides an experience in which users don't use their eyes or hands to interact. So everything you know about visual hierarchy, colors, and using motion to create impact on your user you cannot apply when designing a VUX. The choice of words will influence how people perceive the customer experience you design for them, because there are no accompanying visual clues to guide the user.

Here are some tips:

- Our eyes like patterns. They allow users to complete their action faster. Consistent visual design is good design! However, when it comes to voice, our ears do not like repetition. When designing a voice dialog, **try to use different phrases that signify the same action** (Figure 11-12).

Figure 11-12. *Different phrases signifying the same action*

- **Design for how people talk**, rather than how they type. Users don't use keywords like when they are searching the Internet, for example, "tourist info Zion."

- **Handle errors gracefully**. Avoid error messages that only say that the chatbot didn't hear or understand the user correctly. For example, "I didn't hear you." This causes users to repeat the same phrase that caused the error. Instead, add in information that is more helpful and be as explicit in your directions where possible. Users will quickly lose interest in and be annoyed by talking to a voice that robotically repeats "I'm sorry. I didn't quite catch that," like a broken phone tree menu. Instead, **ask for the missing information only** and **provide a wide variation in potential responses**. Create meaningful error messages to steer the conversation with the user back on track, without being incessantly annoying.

- Outside of what a chatbot is actually saying (the content), its voice reveals a wealth of metainformation to the listener. **Use gender, age, inflection, tone, accent, cadence, and pace** to craft a particular user experience representing the company's brand. The voice makes the chatbot unique, and the tone makes it sound like a human. Using a likable voice is critical for your brand, but there's more to it. Companies that offer a voice UX with a consistent voice and pay careful attention to its tone, build better relationships with their users.

- The tone of the voice is an adaptive reflection of feelings. Think of it in terms of your own voice and tone: You have only one voice, but you probably use a certain tone when you're talking to your friends and an entirely different one when talking to your boss or to customers. You might use an extremely casual tone when you're joking around with a friend, but a serious one when you're consoling someone who's upset. We instinctively adjust our tone in conversation, as an expression of empathy.

- When it comes to audio, you have to make a design decision: "**To record, or not to record?**" :) Do you use a standard voice (i.e., the text-to-speech engine) or do you record custom audio for every response your skill will have?

Always test how well the words flow together and how voice prompts sound to the ear.

- **Be relatable.** Design your VUX so that the chatbot talks **with** your users, not **at** them. Your users need the chatbot to speak concisely to help them understand what information your skill needs and to feel confident about what is happening.

- Given that speech takes precious memory away from other tasks, it is best to keep it minimal. Do not give the user all the information in one go. **Provide only the most relevant information** and then confirm with the user which part to elaborate on.

- Present a list of options to the user by **prioritizing and summarizing the information based on known user preferences**, prior to delivering an answer.

Figure 11-13. *Asking follow-up questions*

- So provoke a more user-oriented voice interaction by **asking relevant follow-up questions** (Figure 11-13) to narrow down the list to the very best options before recommending them. The chatbot can collect the data it needs to provide the best possible answer.

- Finally, keep in mind that users tend to assume that the system can understand beyond its actual capabilities and that very often users are unaware of the available functionalities. Therefor **add help information** whenever the user commits the same mistake twice. Additional to that, whenever relevant to the current situation, the digital assistant can **provide information especially about functions that users have never used before**.

Summary

In this chapter, you have learned how to add sentiment analyses and speech to your Digital Assistant. Sentiment analysis can be implemented in various ways in the context of a chatbot, but typically uses specific, sometimes third-party, APIs dedicated to sentiment analyses. By using a custom component, you can invoke these APIs and work with their result to reply to the user in the correct way.

Exposing your Digital Assistant over voice channels such as Google Home or Amazon Alexa is very straightforward and is based on webhook channels. Once you have your, preferably reusable, webhook code in place, it just a matter of configuration for you Digital Assistant to use these channels. Even though it is very tempting to use voice channels, you must always remember that voice usually requires a completely different design than text channels.

Index

© Luc Bors, Ardhendu Samajdwer, Mascha van Oosterhout 2020
L. Bors et al., *Oracle Digital Assistant*, https://doi.org/10.1007/978-1-4842-5422-6

Printed in the United States
By Bookmasters